Weaving the World

Weaving the World

Simone Weil

on Science,

Mathematics,

and Love

VANCE G. MORGAN

University of Notre Dame Press
Notre Dame, Indiana

Copyright © 2005 by University of Notre Dame
Notre Dame, Indiana 46556
www.undpress.nd.edu
All Rights Reserved

Manufactured in the United States of America

Library of Congress Cataloging-in-Publication Data

Morgan, Vance G.
Weaving the world : Simone Weil on science, mathematics, and love /
Vance G. Morgan.
p.   cm.
Includes bibliographical references and index.
ISBN 0-268-03486-9 (hardcover : alk. paper)
ISBN 0-268-03487-7 (pbk. : alk.paper)
1. Weil, Simone, 1909–1943.   2. Science—Philosophy.
3. Mathematics—Philosophy.   4. Love—Philosophy.   I. Title.
B2430.W474M67   2005
194—dc22
2005019662

∞ *This book is printed on acid-free paper.*

Eternal mathematics . . . is the very stuff
of which the order of the world is woven.

—Simone Weil, The Need for Roots

For Caleb and Justin

Sons are a heritage from the Lord. . . .
Like arrows in the hand of a warrior
are the sons of one's youth.
                    —Psalm 127:3, 4 (RSV)

Contents

Preface

My interest in Simone Weil began over a decade ago when, at a silent retreat, I came across a collection of her letters and essays entitled *Waiting for God.* I was in my second year of teaching philosophy after graduate school and, as many new professors, wondered whether I was having any impact on my students, whether they were learning anything, and whether I knew what I was doing in the first place. I read Weil's essay "Reflections on the Right Use of School Studies with a View to the Love of God," and I was hooked; I read the rest of that beautiful volume in one sitting. The thought and life of this strange and profound woman has, over the past decade, impacted my life intellectually, emotionally, and spiritually in ways I often find difficult to comprehend.

In many ways, Weil and I are unlikely friends. She was a mystic; I am not. She was an agnostic Jew attracted later in life toward Catholic Christianity; I was raised in Protestantism. She tended toward pessimism; I tend toward optimism. Philosophically, her mentor was Plato; my philosophical orientation is guided more by Aristotle. She was a philosophical idealist; I am more of a philosophical realist. Yet she embodies for me what I have always sought as a philosopher, a person who was unafraid to shine the light of her considerable intellectual powers into all corners of her life and reality around her, including finally the mystical experiences that changed her life profoundly in her final years. Simone Weil is a shining example of how reason and faith, intellect and spirit, can and should interact in a person's search for meaning and truth.

Those aware of Simone Weil's life and work are likely to be most familiar with her political activism or her later religious thought. My hope in this book is to bring to the reader's attention a largely unknown aspect of her thought, her lifelong interest in mathematics and science. Her work in science and mathematics not only provides an important doorway into the work of an exceptionally important thinker but also directly reveals many features of the metaphysical foundations that support all of Weil's thought. Finally, Weil's attitudes toward mathematics and science provide a fresh alternative to the predominant naturalistic paradigms of science that tend to bar fruitful interaction between science and disciplines such as philosophy, art, politics, and theology.

This book could not have been written without the support and help of many people. I am grateful to Providence College for granting me a sabbatical semester during the spring of 2002; it was during that semester that the entire first draft of this book was written. My thanks to the members of the American Weil Society, a remarkable group of scholars who welcomed me into their fold with open arms when I was but a neophyte in the things of Weil. Their positive reaction to a paper I presented at the 2001 Weil colloquy at the University of Notre Dame encouraged me to look in more detail at Weil's philosophy of science and mathematics; this book is the result. Special thanks are due to Eric Springsted, John Dunaway, Bernard and Jane Doering, and the late Martin Andic. The comments of two anonymous readers made the final product much better than it could have otherwise been. Finally, I owe an eternal debt of love and gratitude to Jeanne DiPretoro.

Abbreviations

The following abbreviations refer to English editions of works of Simone Weil, which are listed in full in the bibliography.

FLN	*First and Last Notebooks*
FW	*Formative Writings*
GG	*Gravity and Grace*
IC	*Intimations of Christianity Among the Ancient Greeks*
LP	*Lectures on Philosophy*
NB	*The Notebooks of Simone Weil*
NR	*The Need for Roots*
SE	*Selected Essays, 1934–43*
SL	*Seventy Letters*
SN	*On Science, Necessity, and the Love of God*
SW	*Simone Weil*
WG	*Waiting for God*

Kosmos and Crisis

> To restore to science as a whole, for mathematics as well as for psychology
> and sociology, the sense of its origin and veritable destiny as a bridge
> leading toward God—not by diminishing, but by increasing precision in
> demonstration, verification and supposition—that would indeed be a
> task worth accomplishing.
>
> Simone Weil, *NB,* 441

Nobel Prize winner Czeslaw Milosz wrote of Simone Weil that "(her) unique place in the modern world is due to the perfect continuity of her thought. Unlike those who have to reject their past when they become Christians, she developed her ideas from before 1938 even further, introducing more order into them, thanks to the new light."[1] The truth of this claim is confirmed when one considers an area of Weil's thought that has received little attention in the increasing volume of secondary literature pertaining to her life and thought—her lifelong interest in science and mathematics.[2] "Science and Perception in Descartes,"[3] written by the twenty-one-year-old Weil as her capstone thesis for the *agrégation* degree at Ecole Normale Supérieure, introduces themes and concerns related to science and knowledge that form the foundation of a scientific vision that deepens in both richness and complexity over the next thirteen years. The final section of *The Need for Roots,*[4] written months before Weil's death in 1943, includes a call for a radically changed vision of science, rooted in a sacramental conception of reality that alone, in her estimation, can renew hope for humanity in the wake of the tragedies of the first half of the twentieth century. No matter where one looks in the body of Weil's writings between these "bookend" texts, from her published books and essays to her letters and notebooks, one can scarcely read more than a few pages before encountering her insights and reflections on science and mathematics. Accordingly, an extended investigation of these insights will open the way to the thought of one of the twentieth century's great geniuses, of whom Father Perrin nevertheless said that "her soul was incomparably superior to her genius."[5]

As is the case with all of Weil's thought, it is impossible to follow the development of her "philosophy of science" in a simple way; it is as difficult to categorize her reflections on science and mathematics as it is to categorize the woman herself. In Weil's estimation, a proper understanding of the human activity of scientific inquiry requires consideration of where science intersects with history, sociology, politics, epistemology, metaphysics, aesthetics, and religion, to name just a few related investigative activities. In many ways, she considers all of these to be essentially the same activity with different emphases. Furthermore, science for Weil is rooted in the most basic and fundamental human concerns, those of the human body and especially of human labor. An orientation to these complex matters begins with a consideration of some of the various ways in which Weil defines science.

WHAT IS SCIENCE?

In his recent book *Unweaving the Rainbow*, Richard Dawkins quotes the following from a lecture given by the Indian astrophysicist Subrahmanyan Chandrasekhar in 1975:

> This "shuddering before the beautiful," this incredible fact that a discovery motivated by a search after the beautiful in mathematics should find its exact replica in Nature, persuades me to say that beauty is that to which the human mind responds at its deepest and most profound.[6]

The ancient Greeks had a word that encompassed the phenomenon described by Chandrasekhar—*kosmos*. Although this word has come to refer to "the whole world," or "all that exists," as in "cosmology," the study of the origin and structure of the universe, it originally had two basic meanings: order and ornament/adornment. Pythagoras is said to have been the first to apply the word *kosmos* to the universe; according to this use, the world is ornamented with order. In other words, the universe is *beautifully* ordered, its beauty arising from the very ordered structure that lies at its foundations. Calling the world the *kosmos* combines the notions of regularity, tidiness, and arrangement on the one hand with beauty, perfection, and positive moral value on the other.

There are several definitions of science scattered throughout Simone Weil's writings, definitions that on the surface appear to be unrelated to, if not inconsistent with, each other. As we shall see, the metaphysical framework within which she defined science matured in depth and complexity over time. Furthermore, the apparent inconsistencies in the various definitions arise at least partially because they reflect the influences of the Pythagoreans, Plato, Descartes, and Kant, philosophies that are by no means entirely compatible in the context of science and mathematics. When considered closely, however, Weil's definitions provide clues as to how in her thought order, necessity, equilibrium, beauty, truth, and justice are all the proper objects and fruits of scientific inquiry.

Above all, her various definitions of science unfailingly reveal her indebtedness to the ancient Greek notion of science as the investigation of *kosmos* understood as revealing the connectedness of beauty and order. For her, "science, like art, is nothing but a special reflection of the beauty of the world"[7] and "the object of science is the exploration of the beautiful a priori."[8] Most common to her various definitions of science, however, reflecting the influence of Descartes as well as the ancient Greeks, is the concept of *order*. In "Science and Perception in Descartes," she writes simply that "science is nothing but order,"[9] expanded in a later text to "science in all its branches, from mathematics to sociology, has the order of the world for its object."[10] In fact, Weil believes, it is the order of the world, properly considered, that makes scientific study an activity of a special sort. In an early notebook, she quotes her mentor, Alain (Emile Chartier), as having said that "science was born from religion,"[11] writing later that "science—like every human activity—

contains an original, specific way of loving God. And this is its destination, but also its origin."[12] This sacramental conception of scientific activity, heavily indebted to the Pythagoreans and Plato, is of course quite foreign to our contemporary notion of the purpose of scientific activity, primarily that of gaining knowledge of and power over the world. Weil's notion that science is a specific form of religious activity seems, at this point, to indicate that the ultimate object of science is God, pursued by a better and better understanding of the order imposed on matter by God.

In a complex sense, this will turn out to be partially the case. Order, however, is a far more complicated issue in Weil's thought than it originally appears to be, as the following definition of science reveals: "Science is an exploration of all the order that appears in the world, on the scale of our physical and mental organs."[13] This passage reflects the influence of Kant, in that order is identified as a feature not just of the world "out there," but also of the human investigator; order, in other words, arises from the joint activity of the world acting on the knower and the knower imposing structure on the world according to his or her cognitive and physical capacities. Because the order we experience is "on the scale of our physical and mental organs," our scientific inquiries are both structured and limited by our humanness. "Science will always depend largely upon man's intelligence and physique, which are limited and do not become less so with the passing of centuries."[14]

Since science is always enclosed within the parameters of our human abilities and limitations, our common contemporary belief in our unlimited capacity to gain greater knowledge of the world using better and better scientific techniques and equipment is necessarily misguided.

> Science has as its object the study and the theoretical reconstruction of the world—the order of the world in relation to the mental, psychic, and bodily structure of man. Contrary to the naive illusions of certain scholars, neither the use of telescopes and microscopes, nor the employment of most unusual algebraical formulae, nor even a contempt for the principle of noncontradiction will allow it to get beyond the limits of this structure. Moreover it is not desirable that it should.[15]

If the object of science is not to continually increase our mastery of and power over nature, what *is* the ultimate purpose of studying the order of the world? Weil's reflections here connect the primary notions of truth and beauty into scientific activity.

For Weil, the proper attitude in our scientific study of the world is one in which we consider "the order of the world as object of contemplation and of imitation."[16] This places scientific activity in a realm akin to art and religion, in which the order of the world is something to which we are to submit and obey, not overcome. The order of the world, whether a feature of the external world or our fashioning of that world through our own capacities, is a reflection of what Weil frequently calls the Word (*Logos*), the cosmic ordering principle, the knowledge of which is the ultimate goal of all human pursuits. As she indicates in the following sentences from her notebooks, human access to this ordering principle is the primary purpose of science, art, and religion: "Science, art and religion are connected together through the notion of order of the world, which we have completely lost";[17] "Science and art both have one and the same object, which is to experience the reality of the Word, the ordering principle."[18] Hence, to say that "science is an exploration of all the order that appears in the world" connects science to other human investigations of order, such as art and religion, from which science in the contemporary sense is entirely divorced.

Although Weil's later thought will place this sacramental vision of science in an increasingly Christian context, the primary elements of this vision are in place well before the mystical experiences that turned her toward Christianity. The ordering principle, the Word, that is the ultimate object of science is impersonal, revealing itself in those places that are of most concern to human beings. "What is sacred in science is truth; what is sacred in art is beauty. Truth and beauty are impersonal,"[19] and "Justice, truth, and beauty are the image in our world of this impersonal and divine order of the universe."[20] The beauty of the world is the order of the world imposed upon matter by the divine ordering principle and us as human knowers. For Weil, the proper orientation of science is "towards obedience and not towards power,"[21] obedience directed toward the divine order of the world rather than power directed toward overcoming it.[22]

This brief sketch of some of the notions behind Weil's conception of science clearly indicates both just how closely related its metaphysical framework is to that of the Pythagoreans and Plato as well as just how far removed her vision is from the actual activities and energies of contemporary science.

Weil is fully aware of this distance and insists that in many ways the future of humanity depends on bringing science back to its proper orientation to truth and beauty.

THE CRISIS OF CONTEMPORARY SCIENCE

In *The Need for Roots*, Simone Weil argues that in our contemporary world, the authority of scientists and scientific inquiry has replaced, for many, the authority that for many centuries was reserved for religious authorities—in our time, scientists are our "secular priests." She writes that "although science is strictly speaking a matter for specialists only, the prestige which science and savants have acquired over people's minds is immense, and in non-totalitarian countries far and away surpasses any other kind,"[23] adding two pages later that

> So far as the prestige of science is concerned, there are no such people nowadays as unbelievers. That places on savants, and also on philosophers and writers to the extent to which these write about science, a responsibility equal to that which the priests had in the thirteenth century.[24]

If contemporary scientists placed their activities within the sacramental framework described in the previous section, then perhaps the notion of scientists as "secular priests" would be less problematic than it sounds. Not surprisingly, however, Weil finds that contemporary science is something entirely different than "a specific way of loving God"; indeed, it has lost its crucial connection with practical human thought and concerns.

> So long as the intelligence at work in science was simply a more refined form of that which develops the ideas of common sense, there was at least some connection between scientific thought and the rest of human thought, including the thought of the good. But after 1900 even this so indirect connection was lost.[25]

Weil's analysis of how contemporary science has come to be radically different at its core than both ancient and modern/classical (seventeenth- through nineteenth-century) science is complex, and will be a major part of our concerns in the early chapters of this book. At this point, it suffices to consider briefly Weil's claim that although science has in the minds of many contemporary persons taken the place of religion in terms of moral authority, contemporary science has become completely disconnected from what should be its sole concern, namely *truth*. Although scientific techniques and instruments continually become more and more refined, all of the connections to the lives and aspirations of human beings are severed.

> Our science is like a store filled with the most subtle intellectual devices for solving the most complex problems, and yet we are almost incapable of applying the elementary principles of rational thought. In every sphere, we seem to have lost the very elements of intelligence: the ideas of limit, measure, degree, proportion, relation, comparison, contingency, interdependence, interrelation of means and ends.[26]

Scientists, whose sole concern should be the pursuit of truth, have in Weil's analysis over the last century given up on the notion of truth altogether, due at least in part to the influences of quantum theory and the theories of relativity. Indeed, Weil speculates that the irrationality and counter-intuitiveness of these theories actually was attractive to many persons who should have known better: "Men who called themselves philosophers, being weary of reason—no doubt because it is too exacting—rejoiced at the idea of a clash between reason and science; and needless to say their verdict was unfavorable to reason."[27] Once the pursuit of truth is discarded, Weil argues, the remaining vacuum is filled by something frightening.

> So as soon as truth disappears, utility at once takes its place. . . . Everything is oriented towards utility, which nobody thinks of defining. . . . It is as though we had returned to the age of Protagoras and the Sophists, the age when the art of persuasion—whose modern equivalent is advertising slogans, publicity, propaganda meetings, the press, the cinema, and radio—took the place of thought and controlled the fate of cities and accomplished coups d'etat.[28]

When the pursuit of utility replaces the pursuit of truth, power becomes the dominant driving force behind scientific investigation. Science, once freed from elements such as truth and beauty which anchor scientific inquiry to the whole of human life and values, ceases to be science in a traditional sense and is transformed into the servant of technology, mechanization, and power. In a striking passage toward the end of *The Need for Roots*, Weil quotes some lines from Hitler's *Mein Kampf* in which Hitler observes that in the natural world "force reigns everywhere and supreme over weakness, which it either compels to serve it docilely or else crushes out of existence."[29] In such a world, human beings are naive if they expect that they are exempt from these laws of force—human struggle is equally a matter of power dominating over weakness. As Weil reminds us, "Hitler's life is nothing but the putting into practice of that conclusion."[30]

In 1943, as now, it was easy to view Hitler and others like him as monstrous aberrations. Weil cautions us, however, against such a facile conclusion.

> These lines [from *Mein Kampf*] express in faultless fashion the only conclusion that can reasonably be drawn from the conception of the world contained in our science. . . . Who can reproach him for having put into practice what he thought he recognized to be the truth? Those who, having in themselves the foundations of the same belief, haven't embraced it consciously and haven't translated it into acts, have only escaped being criminals thanks to the want of a certain sort of courage which he possesses.[31]

To the extent that our science tells us that we exist in an irrational world dominated by force, to that same extent truth, beauty, and justice become mere words with no content. Scientific paradigms are part of the framework within which human beings think and act; Weil's discussions of such paradigms are of far more than mere historical or academic interest if she is even at least partially correct about the devastating problems underlying the paradigms of contemporary science. In a 1937 letter to Jean Pasternak, she writes that

> I believe as you do that science is entering a period of crisis graver than that of the fifth century, which is accompanied, as then, by a moral crisis and a subservience to purely political values, in other

words, to power. The new phenomenon of the totalitarian State makes this crisis infinitely formidable, and may turn it into a death-agony.[32]

In 1943, she has become even more convinced of the truth of her belief. "The modern conception of science is responsible, as is that of history and that of art, for the monstrous conditions under which we live, and will, in its turn, have to be transformed, before we can hope to see the dawn of a better civilization."[33] Many of the issues to be discussed in the pages to come relate directly to this claim. How did the modern conception of science arise? What alternative paradigms are available? Is her call for a transformed science leading to a transformed humanity the least bit practical, or is it merely expressive of hopes most appropriate to mystics and idealists?

Weil tells us that hope arises first because even if the logical conclusion of our contemporary scientific paradigms has no place for truth, beauty, and justice, that very conclusion indicates that our scientific paradigms must be insufficient. This is because we know, as experiencing human beings, that truth, beauty, and justice *do* exist.

> If pure good were never capable of producing on this earth true greatness in art, science, theoretical speculation, public enterprise, if in all these spheres there were only false greatness, if in all these spheres everything were despicable, and consequently condemnable, there would be no hope at all for the affairs of this world; no possible illumination of this world by the other one.
>
> But it is not so.[34]
>
> Where force is absolutely sovereign, justice is absolutely unreal. Yet justice cannot be that. We know it experimentally. It is real enough in the hearts of men. The structure of the human heart is just as much of a reality as any other in this universe, neither more nor less of a reality than the trajectory of a planet. . . . If justice is inerasable from the heart of Man, it must have a reality in this world. It is science, then, which is mistaken.[35]

Weil will seek for new scientific paradigms in various places, but first and foremost in the scientific energies of the ancient Greeks, for whom, as described above, science, religion, and art were unified as investigations into the divinely imposed order of the world.

Above all else, humanity's spiritual thirst must be addressed by science in a way that it has not been for centuries. In a 1941 letter to Déodat Roché, Weil writes that

> We are living at a time when most people feel, confusedly but keenly, that what was called enlightenment in the eighteenth century, including the sciences, provides an insufficient spiritual diet; but this feeling is now leading humanity into the darkest paths. There is an urgent need to refer back to those great epochs which favored the kind of spiritual life of which all that is most precious in science and art is no more than a somewhat imperfect reflection.[36]

The urgency of this need should not blind us to the fact that, although possible, satisfying the need will be difficult, requiring a wholesale transformation of human orientation.

> Today it is not the fate of Greece but of the entire world that is at stake. And we have no Socrates or Plato or Eudoxus, no Pythagorean tradition, and no teaching of the Mysteries. We have the Christian tradition, but it can do nothing for us unless it comes alive in us again.[37]

Science and Work

The secret of the human condition is that equilibrium between man and the surrounding forces of nature "which infinitely surpass him" cannot be achieved by inaction; it is only achieved in the action by which man recreates his own life: that is to say, by work.

—Simone Weil, *FLN*, 18

Simone Weil was born into a world in the throes of radical scientific change. As Weil wrote toward the end of her life, "To us, men of the West, a very strange thing happened at the turn of this century; without noticing it, we lost science, or at least the thing that had been called by that name for the last four centuries. What we now have in place of it is something different, radically different, and we don't know what it is."[1] The theories that caused the rupture between the traditional conception of science and the unknown "something" that took its place were Einstein's theories of relativity and quantum theory. Weil came to find these theories disturbing on many levels and spent a great deal of time in her later years reflecting on their implications, not only for science, but for the entire scope of our understanding of ourselves. Even her earliest writings, however, reveal a deep sensitivity to the important role that our understanding of the nature of science plays in our vision of the human person and its place in the *kosmos,* as well as to the threat that new paradigms of science raise to this vision.

This chapter focuses on these seminal reflections, in particular on "Science and Perception in Descartes"—Weil's dissertation written for the *diplome d'études supérieures* at the Ecole Normale—and *Lectures on Philosophy.* The former text is a reexamination of the relationship between Descartes, the "Father of Modern Philosophy," and modern science, in which Weil develops an alternative reading of familiar texts from Descartes which, if supportable, lays the foundation for a very different model of the human person and reality than the dualist and mechanist vision traditionally attributed to Cartesian thought. The later text consists of notes of Weil's lectures as a philosophy instructor at the Roanne *lycée* in 1933–34, taken by one of her students, Anne Reynaud-Guérithault. These texts are of general value and interest in that they reveal Weil the philosopher at work; she was trained as a philosopher first, and her considerable skill as a philosopher never failed her either as a social activist or in her later religious thought. More specifically, these texts show that from the outset, Weil envisioned science as a practical activity at the core of human life. This scientific "rootedness" only deepened in successive years under the increasing influence of the Pythagoreans and Christianity.

HISTORY'S GREATEST MOMENT

In Weil's introduction to "Science and Perception in Descartes," we find the following remarkable passage:

> The advent of the geometer Thales was history's greatest moment, a moment that is repeated in an individual's life, for Thales is reborn for each new generation of students. Until the time of Thales, humanity had proceeded only by trial and error and guesswork; from the moment that Thales discovered geometry . . . humanity knew.[2]

Thales is considered not only to be the first geometer but also the first philosopher in the modern sense of the term. He is credited with taking the first important steps toward demythologizing explanations of natural phenomena, explaining earthquakes, for instance, by his belief that the earth rests on water rather than the traditional explanation that they were due to the activity of Poseidon. Such explanations involved hypothesizing an unobserved natural state of affairs to account for observed natural phenomena; by generating such explanations, Thales was making an intellectual move which has remained a principal part of scientific thinking to this day. Thales is also credited with discovering several mathematical and geometrical theorems, which he used to demonstrate the distance of ships at sea from the shore. Most famously, he measured the height of the pyramids by comparing the ratio of their height to their shadow with the ratio of a man's height to his shadow.

Why does Weil believe that "the advent of the geometer Thales was history's greatest moment"? Because in her estimation he provides the first historical evidence of the possibility of science, of the possibility of explaining the hitherto unexplainable by the use of rational, methodical procedures. Thales' discovery of geometry established that certainty is possible, placing indubitable knowledge within the reach of all human beings who are willing to think carefully and methodically. Weil suggests that while human beings have always sought "a higher, more certain knowledge" than the "immediate interpretation of sensations,"[3] before Thales, common humanity believed such certainty to be available only to an elite whom they made their priests and kings. Thales took the beginning steps in the direction of science by showing that the natural world can be understood and productively engaged by all thoughtful and enterprising human beings. In *Lectures on Philosophy*, Weil provides her students with a simple but effective illustration of the revolution that began with Thales.

> One gets the impression that there is some evil power in things when they present us with obstacles that we cannot overcome. The catastrophes which make us lose our heads, lead us to say: "Is this a dream?" If, now, one supposes that the same men, faced with the same blocks of stone, instead of reacting blindly, begin to reflect about the situation in an ordered way and use a lever, everything changes. . . . It is this idea that overcomes all the evil force in the world.[4]

The discovery that geometry and mathematics provide the key to understanding natural phenomena would seem to place such understanding within the reach of all human beings, since the discovery demonstrates that indubitable knowledge is directly linked to commonsense perception. No longer is certainty the exclusive property of an elite. Still, it may be that scientific knowledge is rooted in cognitive capacities that are possessed only by a new elite of scientific intelligentsia. The advent of science, Weil argues, presents humanity with a dilemma.

> This revolution . . . overthrew the authority of the priests. But how did it do so? What did it bring us in its place? . . . Did it replace tyrannical priests who ruled by means of the tricks of religion with true priests who exercise a legitimate authority because they really have access to the intelligible world? . . . Or, on the contrary, did this revolution replace inequality with equality by teaching us that the realm of pure thought is the sensible world itself, that this quasi-divine knowledge that religions sensed is only a chimera, or rather, that it is nothing but ordinary thought? Nothing is harder to know, and at the same time is more important for every man to know. For it is a matter of nothing less than knowing whether I ought to make the conduct of my life subject to the authority of scientific thinkers, or solely to the light of my own reason.[5]

The issue of collective authority, whether religious, political, or scientific, is of great concern to Weil throughout her life.[6] To the extent that scientific progress engenders an elite, it wanders dangerously from its roots in common human activities. Yet, Weil argues, that is precisely the track that science in the Western tradition has taken, beginning, interestingly enough, with the Greek tradition within which science first arose.

Although Thales' own application of geometry to practical problem solving rooted science in common human experience, the Greeks over time came to think of geometry very differently. Geometry became an activity closer to religion than practical problem solving. Weil reminds us that for Plato,

> although the geometer makes use of figures, these figures are not the object of geometry, but only the occasion for reasoning about the

ideal straight line, the ideal triangle, the ideal circle. . . . The philosophers of the Platonic school reduced . . . the whole of what is perceived to a fabric of appearances, and forbade the search for wisdom to anyone who was not a geometer.[7]

As one climbs Plato's Divided Line, one must rise above the physical, practical world if one is to encounter fully the truth of mathematical and geometrical relationships. The world of such relationships is reality, while the world of matter and the body is mere appearance. From the outset, then, the relationship of mathematical knowledge to practical experience is ambiguous.

When one turns one's attention to modern science, the abyss between science and practicality has become even greater; "it is truly another realm of thought that modern science brings us."[8] As Weil wrote in "Human Personality" late in life, "In our days, when writers and scientists have so oddly usurped the place of priests, the public acknowledges, with a totally unjustified docility, that the artistic and scientific faculties are sacred. This is generally held to be self-evident, though it is very far from being so."[9] In "Science and Perception in Descartes," Weil speculates that if Thales himself returned to see how Western science has built upon his insights, he "would feel like a son of the soil compared to our scientists."[10] This is not because of the technical advances and invention of better instruments of observation over the past twenty-five hundred years. Rather, Thales would be amazed to find that contemporary science has become abstract to the point that astronomy books make little mention of stars, books on heat contain no definition of heat or explanation of how it is propagated, and books concerning the nature of the physical world have no physical or mechanical models in them. Thales would think that his discovery of the applicability of geometry to the real world has been forgotten.

> Science, which in the time of the Greeks was the science of numbers, geometrical figures, and machines, now seems to be solely the science of pure relationships. Ordinary thinking, which it seems Thales at least utilized, even if he did not confine himself to it, is now clearly scorned. Common-sense notions such as three-dimensional space and the postulates of Euclidean geometry have

been discarded. Certain theories do not hesitate to speak about curved space, or to put a measurable speed next to an infinite one. Speculative theories about the nature of matter are given free rein, in the attempt to interpret this or that result of our physics without the slightest concern for what matter may mean to ordinary men who touch it with their hands.[11]

What has happened? According to Weil, scientists now express the geometrical truths of the Greeks in the mathematical language of algebra, to the extent that "scientists . . . no longer admit anything into science except the most abstract form of reasoning, expressed in a suitable language by means of algebraic signs."[12] Algebra is a mathematical language that facilitates the comparison of natural phenomena. It is an exceptionally powerful tool for analyzing such phenomena so long as it is not forgotten that "the value of a language is to be found in a relationship between language and something else."[13] In Weil's estimation, under the use of contemporary scientists algebra has become a language entirely disconnected from the representation of anything; hence, the results of contemporary science cannot be pictured or intuited in any reasonable or commonsensical manner.[14] "The mathematician lives in a universe apart, whose objects are signs. The relation between sign and thing signified no longer exists; the play of interchange between signs develops of itself and for itself."[15] For this reason, "scientists have thus indeed become the successors of the priests of the old theocracies."[16]

Weil illustrates this dynamic in *Lectures on Philosophy* with a brief discussion of the work of the German physicist Heinrich Hertz, the discoverer of Hertzian waves and the photoelectric effect. After his experiments with electricity, Hertz was able to reduce his analyses to algebraic formulae that were remarkably similar to the algebraic formulae that represent the activity of light. From this similarity of algebraic formulae, Hertz concluded that light is an electromagnetic phenomenon. Weil comments that "there is something remarkable in this kind of hypothesis: one does not know the nature of electricity."[17] In other words, Hertz's hypothesis tells us nothing about either the nature of electricity or light other than that their effects can be represented by similar algebraic formulae. "Hypotheses of this kind are very obscure . . . they cannot be represented in the imagination . . . They are independent of mechanical representation."[18] In this case, algebra has become a language with no connection to anything other than itself; "in this kind of hypothesis, algebra takes the place of human thought."[19]

Weil's overall project in "Science and Perception in Descartes" is to investigate whether the separation of scientific thinking from ordinary thinking flows from the nature of science itself or whether it has come about as a result of various scientific and philosophical prejudices that are actually at odds with the true activity of science. Much of great importance hangs in the balance, because "if . . . the physicist chooses to frame suppositions, laws, and conceptions incompatible with the ideas which are common to all, he will alienate himself not only from the peasant but first of all from himself, from all that part of himself in which he resembles the peasant."[20] For Weil, the best way to address the above problem is not to return back to Thales, but rather "we must go back . . . to the double revolution through which physics became an application of mathematics and geometry became algebra: in other words, to Descartes."[21]

CARTESIAN CONTRADICTIONS

The traditional interpretation of Descartes' philosophy identifies Descartes as the originator of modern science, instituting the separation of scientific investigation from the reach of the senses that is characteristic of science today. Indeed, Weil argues, "from its very origin [with Descartes] modern science has been, albeit in a less-developed form, what it is now."[22] The basic features of this interpretation are familiar. Descartes' dualistic philosophy sharply separates mind from matter, placing certainty within the reach only of mental activity untainted by the senses or the imagination. Matter is mere extension, to be investigated according to mathematical relationships best described in algebraic formulae. Science becomes abstracted from the material world familiar through the senses, hence scientific descriptions become entirely disconnected from everyday human experience. Although Descartes is often regarded, in a manner bordering on caricature, in contemporary thought as the foundation of everything that is oppressive and limiting in modern science and culture because of his radical division of mind from matter, there is no doubt that the modern Western conception of science is largely indebted to his philosophical and scientific vision.

In part 1 of "Science and Perception in Descartes," Weil begins by demonstrating that the traditional interpretation of Cartesian philosophy is strongly supported by textual evidence from Descartes' body of work. Not only does Descartes begin his philosophical enterprise by doubting the evidence of the senses, he never gives sensory input back to us as evidence that can be fully trusted even at the end of his investigations. In terms of science, "the aim of Cartesian physics is to replace the things that we sense with things that we can understand only."[23] The entire thrust of Descartes' scientific vision is to put one's trust in reason alone, even in the face of directly contrary sensory evidence. Descartes' a priori method is intended to establish the nature of the world around us apart from the evidence of the senses; Weil illustrates this conviction with the following "almost insolent remark" from Descartes' *Principles of Philosophy*: "The proofs of all this are so certain that, even if experience might seem to show us the contrary, we would nevertheless be obliged to put more faith in our reason than in our senses."[24]

It is when one considers the fact that Cartesian physics is entirely a matter of investigating geometrical relationships that one can see why Thales would not recognize modern science as anything closely related to his own discoveries. Although Thales and Greek geometricians touched off the development of science by showing that the natural world can be engaged through the application of mathematical and geometrical properties, their investigations were always intuitive and always preserved something that was specific to the kind of figure that they were studying. For instance, Thales' "proof" that a circle is bisected by its diameter may well have involved folding or cutting a drawn circle and showing that the pieces match.[25] Although this does contain the germ of mathematical proof, such a procedure would hardly count as a legitimate proof in modern science. Thales' discovery concerning the circle is still rooted in the physical circle itself.

It was Descartes who "liberated" science and mathematics from the chains of the physical world. "Descartes was the first to understand that the sole object of science is the quantities to be measured, or rather the ratios that determine this measurement—ratios that, in geometry, are found only as it were hidden in figures, in the same way that, for example, ratios can be hidden in movements."[26] From Descartes on, mathematicians and scientists considered that "figures were no more than givens that posited ratios of quantity; all that was left was to apply arithmetical signs to ratios of this new sort. . . . In short, from 1637 [the date of the publication of Descartes' *Geometry*] on, the essence of the circle, as Spinoza would later say, was no longer circular."[27] Descartes' great insight is that the order exhibited by the mathematical method, an order that produces indubitable and certain results, can be applied to any investigation whatsoever. Since this method is imposed by the mind on the world, its origin and application must be a priori, untainted by sense experience. With this in mind, Weil argues,

> The idea that we can form of Descartes as the founder of modern science thus seems complete. Classical geometry was still as it were glued to the earth; he set it loose, and was like a second Thales in relation to Thales. He transferred the knowledge of nature from the realm of the senses to the realm of reason. He thus rid our thought of imagination, and modern scientists, who have applied mathematical analysis directly to all the objects that can be studied in that way, are his true successors.[28]

It would seem from textual evidence that the traditional dualistic, mechanistic interpretation of Descartes is more than warranted. His philosophy does indeed lay the foundation for a science that is entirely disconnected from sense perception, and hence from common sense and understanding itself.

This would be the end of the story, except for the fact that a more careful reading of the entire Cartesian corpus brings to light numerous passages that conflict directly with the traditional interpretation. It is clear from these texts that Descartes is intensely interested in science, not for its own sake, but for its practical applications. Although the metaphysical foundations of Cartesian science are purely rational, Descartes expressed in the *Discourse on Method* that the purpose of his physics is

> to arrive at knowledge that is very useful in life . . . in place of the speculative philosophy that is taught in the schools, we can find a practical philosophy, by means of which, knowing the power and the actions of fire, water, air, the stars, the heavens, and all the other bodies that surround us as clearly as we know the various trades of

our craftsmen, we could, in the same way, put these things to all the
uses to which they are appropriate, and thus render ourselves as it
were masters and possessors of nature.[29]

Such passages indicate that for Descartes, "science seems to be regarded not
as the means of satisfying our curiosity about nature, but as a method for
taking possession of it."[30] The key to Descartes' philosophical system, the
Cartesian method, is simply the application of order to the object of one's
investigation, an order exhibited by the precision of mathematical investigation. The purpose of mathematics is not the esoteric investigation of abstract
relationships. Descartes makes clear in *Rules for the Direction of the Mind*
that he has little interest in science and mathematics "for their own sakes";
geometrical and algebraic proofs are nothing but the "outer covering" of real
science.

> In truth, nothing is more useless than to be so occupied with empty
> numbers and imaginary figures that we seem to be willing to find
> pleasure in knowledge of such trifles. . . . Nevertheless, whoever will
> have considered my thought carefully will easily understand that I
> am not thinking here of ordinary Mathematics, but am expounding
> another science, of which [these illustrations] are the outer covering
> rather than the parts.[31]

Since the order exhibited by mathematics and the Cartesian method is
a matter of simply directing one's reason well, Descartes is a true egalitarian
with respect to what scientific investigation demands of the human knower.
The proper, orderly use of the human mind is within the reach of all human
knowers in all human investigations; since "all the sciences are nothing other
than human wisdom . . . it is not necessary for minds to be confined by any
limits."[32] If science is nothing but the orderly use of right reasoning, Descartes' conception of science has no room for a scientific "elite" or "intelligentsia." As Weil summarizes,

> not only does Descartes regard every mind, as soon as it makes a
> serious effort to think properly, as equal to the greatest genius, but

he finds the human mind even in the most ordinary thinking. There is, in his eyes, a common wisdom—a wisdom that is to the mind what the eyes are to the body—much closer to authentic philosophy than is the kind of thinking that study produces.[33]

For Descartes, then, mathematics rules science in a way quite distinct from the way in which algebra rules contemporary science. In Descartes' physics, mathematics is not just a convenient language governing the expression of abstract relationships; as Weil writes, "in Descartes, geometry is itself a physics . . . which proposes for questions the explanation of natural phenomena."[34] Algebra itself, with which Descartes replaced geometry, is rooted in the real world. In Descartes we find both extreme idealism and realism, contradictory energies that Descartes himself believed to be "not only reconcilable but correlative."[35] Weil's own lifelong concern with science includes continuing attempts to conceive of a metaphysical framework that can satisfactorily accommodate a scientific vision sensitive to both of these apparently incompatible energies.

One of the reasons that traditional interpreters have tended to downplay or ignore the realist passages in Descartes is that it seems entirely inconsistent to be both an idealist and a realist with respect to anything, particularly science. How is it that "this geometry, so ethereal that it seems to disdain figures, proves to be substantial enough to constitute a physics"?[36] Weil points out that Descartes himself provides ample clues to how he himself supposed that these divergent tendencies can be reconciled. In a letter to Princess Elizabeth of Bohemia, he writes that "the study of mathematics, which exercises mainly the imagination in the consideration of shapes and motions, accustoms us to form very distinct notions of body."[37] By the imagination, mathematics hooks us into the material world of the body. Weil will explore the imagination in a later part of "Science and Perception in Descartes"; at this point, she suggests that for the imagination to operate, more than abstract ideas and relationships are required. "Since the mind in geometry makes use of the imagination, it does not handle empty ideas. It grasps something."[38] We shall see that, in Weil's estimation, it is this "grasping" that constitutes the link between the mind and matter, as well as the union of idealism and realism.

Weil thus concludes that

> Cartesian physics is far more packed with matter than is ordinarily
> thought. . . . It is so bound to the imagination, so joined to the human
> body, so close to the most common labors, that one may be initiated
> into it by studying the easiest and simplest crafts; especially those
> that are the most subject to order, like that of weavers, embroiderers,
> or lacemakers.[39]

There is no denying the fact, however, that Descartes himself never successfully united the idealistic and realistic strands of his philosophy. Commentators and interpreters have focused almost exclusively on the dualistic,
rationalistic Descartes and the mechanistic science that is compatible with
these aspects of his philosophy. Weil, however, believes that there is sufficient material in Descartes' works to justify investigating whether a different
type of science, one less abstract and more rooted in common human activities, might have arisen if more attention had been paid to the obvious interest
that Descartes paid to the real, practical world in his considerations.

At this point Weil suggests a procedure that Descartes himself would
have approved of. Rather than continue developing conflicting interpreta-
tions of Descartes' philosophy from the outside, "every commentator must
become, at least for a time, a Cartesian."[40] This requires universal doubt followed by orderly examination, "without believing in anything except one's
own thought insofar as it is clear and distinct, and without trusting the
authority of anyone, even Descartes, in the least."[41]

> Just as Descartes, in order to form accurate ideas about the world in
> which we live, imagined another world that would begin with a sort
> of chaos and in which everything would be ruled by figure and move-
> ment, so let us imagine another Descartes, a Descartes brought
> back to life. This new Descartes would have at first neither genius,
> nor knowledge of mathematics and physics, nor force of style; he
> would have in common with him only the fact of being a human
> being and of having resolved to believe only in himself. According to
> Cartesian doctrine, that is enough. . . . So let us listen to this ficti-
> tious thinker.[42]

Descartes Brought Back to Life

Part 2 of "Science and Perception in Descartes," in which Weil "becomes a Cartesian," is by no means a simple imitation of Descartes' own procedure. The overarching problem of both exercises is to discover how, from a cognitive slate wiped as clean as possible by systematic doubt, the human world can be constructed with certainty and indubitability. Descartes' procedure, of course, is to first doubt everything that comes through the senses in order to establish what can be known with certainty by the mind alone. Weil begins her Cartesian exercise in a phenomenological reverie, a description of a stream of undifferentiated sensations, the most basic component of which is the awareness of self and world.

> We are living beings; our thinking is accompanied by pleasure or pain. I am in the world; that is, I feel that I am subject to some external thing that I feel is more or less subject to me in return. Depending on whether I feel this external thing submitting to me, or myself being subject to it, I feel pleasure or pain. . . . Pleasure and pain are mixed with one another. . . . This feeling with its shadings of pleasure and pain, which is the only thing I can experience, is thus the fabric of the world. It is all that I can say about the world.[43]

These primordial feelings are the "medium" through which the world appears to consciousness. How does this essentially undifferentiated flow of sensations become ordered into the distinction between "I" and the "world" that confronts me?[44]

In addition to sensations of pleasure and pain, other ideas that do not give rise to pleasure and pain are present in this beginning state. For instance, this "world of ideas" contains arithmetical propositions, which contain a certain irresistibility independent of my wishes or desires, limiting the manners in which my sensations present themselves. "My existence manifests itself to me through the medium of appearances, but it can appear to me only in certain ways; there are ways of appearing that do not define anything that can appear to me."[45] For instance, two pairs of oranges can only add up to four oranges, not five, regardless of my sensations. However, neither pain- or

pleasure-causing sensations nor more abstract ideas are helpful in distinguishing between myself and world. If I ask why my experience occurs in
regular and predictable patterns, the only answer available at this point is
"that's the way I am. . . . What I call the world of ideas is no less chaos than
the world of sensations. Ideas impose their ways of being on me, take hold of
me, escape me."[46] The apparent cohesion, order, and predictability of my sensations and ideas are, at this point, entirely unexplainable—they are chance
features of my conscious experience.

> Chance is clothed, disguised, in blue, in gray, in light, in hardness
> and softness, cold and heat, in the straight line and in the curve, in
> triangles, in circles, in numbers; chance is anything and everything.
> I am never conscious of anything except the trappings of chance;
> and this very thought, insofar as I am conscious of it, is chance.
> There is nothing else.[47]

Even though "I never know what it is that I am conscious of; I know
only that I am conscious of it,"[48] there is something else after all. Consciousness itself is marked by the power of thought, even if there is no way of knowing what causes the ideas and sensations thought refers to. The power of
thought primarily expresses itself through doubting, through the capacity to
either believe or refuse to believe in the truth of what is present to thought.
Thoughts, illusory or not, "need me in order to be thoughts."

> What do they [my thoughts] borrow from me? Belief. It is I who
> think these things that produce illusion, and whether I think of
> them as certainties or as illusions, the spell that they cast over me
> remains intact. The power that I exercise over my own belief is not
> an illusion; it is through this power that I know that I think. . . . And
> through this power of thinking—which so far is revealed to me only
> by the power of doubting—I know that I am. I have power, therefore
> I am.[49]

At first glance, Weil's *je puis, donc je suis* looks remarkably similar to Descartes' more famous *je pense, donc je suis*. The subtle difference between
them, however, is crucial. The "I" whose existence is established in Weil's

argument is not a substantial subject of consciousness (as in Descartes); rather, Weil's focus is on the *activity* of thought itself.

> To exist, to think, to know are only aspects of a single reality: to be able to do something. . . . As for knowing my own being, what I am is defined by what I can do. So there is one thing I can know: myself. And I cannot know anything else. To know is to know what I can do; and I know to the degree that I substitute "to act" and "to be acted upon" for "to enjoy," "to suffer," "to feel," and "to imagine." In this way I transform illusion into certainty and chance into necessity."[50]

Weil's attention is on the fact that thought is the active element in all action; in essence, as Winch says, for Weil the word "I" "functions as the purely grammatical subject of verbs of activity."[51] By focusing exclusively on the activity of thought rather than on the subject doing the thinking, Weil avoids the familiar dualistic problems created by Descartes' sharp separation of mind and body. By establishing the power to act as the first element of certainty in her Cartesian exercise, Weil reveals a philosophical commitment to the connection between thinker and world that remains with her throughout her life.

Since I am defined by what I can do, an investigation of the limits of my control over my thoughts and actions will at the same time be an investigation of the boundary between, or intersection of, myself and the world. Although power in itself is infinite, it is clear that the power I have over my thoughts is limited. This, in itself, establishes the existence of something other than myself.

> All real power is infinite. If nothing but me exists, nothing exists except this absolute power. . . . Although this power that I possess is by nature infinite, it has some limitations that I must recognize. My sovereignty over myself, which is absolute as long as I want only to suspend my thought, disappears as soon as it is a matter of giving myself something to think about. . . . Therefore, something other than myself exists. Since no power is limited by itself, it is enough for me to know that my power is not absolute to know that my existence is not the only existence. What is this other existence?[52]

This "other existence" stamps its imprint on me, limiting my power, yet leaves me with the freedom to place my own imprint on it. As long as I submit to what is presented to me on the level of pleasure and pain, I am at the whim of my sensations. If, however, I disengage myself from the pleasure/pain reaction and seek to establish what power I have to grasp this "other existence," I can begin to define the limits of my power and of myself.

> However much the other existence has power over me through the intermediary of my thoughts, so much, through the same intermediary, do I have power over it. Consequently, although I cannot create a single one of my thoughts, all of them—from dreams, desires, and passions to reasoned arguments—are, to the extent that they are subject to me, signs of myself; to the extent that they are not subject to me, signs of the other existence.[53]

The question now becomes "what must I do to learn more than I know at present?"[54] The type of mental activity that established *je puis, donc je suis* is not helpful in grasping this other existence and my relationship to it, since it required disengaging from precisely what now needs to be grasped. Weil proposes, in a most non-Cartesian manner, that from the start expectations of certainty and precise clarity are misplaced when investigating the intersection of self and world. "The knowledge by which I grasp something other than myself is entirely different; there is no longer any question of asking for something to be made clear."[55] Since neither abstract cognitive activity nor undifferentiated sense impressions can reveal the nature of the mutual interaction of self and world, Weil proposes that "I will go to this third, ambiguous being that is a composite of myself and the world acting on each other . . . I will name this bond the imagination."[56] The imagination is "the knot of action and reaction that attaches me to the world"; only by turning my attention to the imagination will I be able to discover "what is my portion and what it is that which resists me."[57]

In *Lectures on Philosophy,* Weil expresses the intimate nature of our relationship to the world in the following manner: "The very nature of the relationship between ourselves and what is external to us, a relationship which consists in a reaction, a reflex, is our perception of the external world. Perception of nature, pure and simple, is a sort of dance; it is this dance that makes perception possible for us."[58] The order and flow of a dance is attributable to the give and take between the dance partners, alternatively active and passive. The imagination has both aspects. It reveals the presence of an external world via sensations that I can neither control nor understand. Uninfluenced by the mind, the imagination transforms sense impressions into images that impinge upon me and make me subject, through pain and pleasure, to the world. When, however, the imagination serves as the passageway into the world for the mind, it also reveals the mind's capacity to grasp the world. When the mind uses the imagination to grasp the world, it transforms the world from something to be feared and to dread into an obstacle to be overcome.

How is this accomplished? We have arrived at not only the heart of the human being's engagement with the world, but also the source of science. How does the world become subject to the influence of the mind via the imagination? Weil returns here to her earlier observation that some of her ideas are distinctly different from pain- and pleasure-causing sensations. These ideas, such as number and sequence, serve as an ordering tool for the mind.

> I come across an idea of another kind, one that does not impose itself on me, that exists only through an act of my attention, and that I cannot change; it is, like the "I think, therefore I am," transparent and invincible. The idea of number and ideas like it replace the random changes to which other ideas are subject with an orderly progression that originates in themselves. They are primarily useful for kinds of reasoning that, although clear and immutable like them, seem to force me to accept what they present almost in the same way the senses do, and as it were throw truths at me.[59]

By the use of number and sequence, the mind orders the world by methodically applying elementary structures such as arithmetic and geometry to it. This is how "the mind arms itself against chance." "Its passions will give way to a will that, despite the condition to which the world reduces it, will imprint itself directly in things, provided that what it wills is only the kind of change that it is capable of bringing about, and that constitutes its mastery over the world."[60] We turn the world from something to which we are entirely subject to something we can indirectly influence by means of the imposition of order. The activity of ordering and sequencing is unlimited, and in a real sense makes the mind the equal of the world. "Thus, at least to the extent that the world is subject to my action, order gives me the power to hold the world as a whole before my eyes, to examine it, and to be assured that it in no way surpasses my thought."[61]

Weil is clear that the order, precision, and structure of the dance between self and world is a feature of the mind, not of the world, yet it is this very grasping of the world by means of order and method that constitutes our active contribution to the dance. "What marks off the 'self' is method; it has no other source than ourselves: it is when we really employ method that we really begin to exist. . . . In action that has method about it, we ourselves act, since it is we ourselves who found the method: we really act because what is unforeseen presents itself to us."[62] Having intelligible thoughts about the world requires that I experience things as ordered in a definite series. This order is not something experientially given, but is something I construct in reacting methodically to the situations confronting me. "Number is not something that we get from the world; we ourselves, and no one else, are the authors of the series of numbers: for example, the world, in a storm, is not going to provide us with 1 grain, then 2, then 3 grains of sand."[63]

In "Science and Perception in Descartes," Weil speculates concerning how this ordering occurs. She suggests that the simplest of her ideas is the idea of straight line motion. By the continual application of this basic idea to the thoughts produced by sensation and the imagination, she arrives at geometrical forms of increasing complexity—the oblique line, the circle, the ellipse. These forms correspond in their increasing complexity to the series of numbers. This entire structure of ideas is in place prior to turning toward the world, providing her with "the means to replace the ideas that the deceptive imagination makes me read in sensations."[64] By supposing "that, in opposition to the single movement that I control, the world offers an infinitely complex movement that is to motion in a straight line what the number the mathematicians call infinity is to the number one," Weil proposes that she "will suppose there is an indefinite quantity of simple motions in the world, and I will define each of them, like my own, by a straight line." By this procedure, "I can analyze the world endlessly."[65]

It could be argued at this point that even though Weil has attempted to avoid Descartes' radical dualism, her attempt to impose order on the world in this manner is flawed in that it still splits self from world too sharply, leaving no room for an account of the important role of the human body and social contexts in the construction of human knowledge. Weil herself appears to have recognized this problem. In *Lectures on Philosophy*, four years after "Science and Perception in Descartes," her discussion of how knowledge of self and world is constructed is rooted from the start in the human body. She notes that even in non-cognitive reactions to stimuli such as reflexes, the body reacts in an ordered manner, an order more primal and basic than that investigated in "Science and Perception in Descartes." "The body classifies things in the world before there is any thought . . . so, from the very fact that we have a body, the world is ordered for it; it is arranged in order in relation to the body's reactions. . . . When we are on the point of giving birth to thought, it comes to birth in a world that already is ordered."[66] Later, Weil also investigates the important role of language in ordering the human world. In short, her account in *Lectures on Philosophy* is more complex and "materialistic" than in "Science and Perception in Descartes." We do not discover and apply a preexistent necessary order in thought; we participate in a material and social world in which thought occurs.

It is impossible to tell whether the mathematical physics produced by applying geometrical and algebraic analysis to sensation brings me closer to "the way things are," in that from the start this physics is a dual product of mind and world. To reduce the complexity of sensation to mathematically analyzable extension is a "neat trick," one might say, but to what extent can it harness the continuing fantasies of the imagination?

> As long as the wild imagination makes me think that I see the most fantastic things in sensations and makes me conjure up a god in every thought, it is useless for me to try to counter this eloquent madness with the simple supposition that what is actually signified by sensations is extension. . . . I am always a dual being, on the one hand a passive being who is subject to the world, and on the other an active being who has a grasp on it; geometry and physics help me to conceive how these two beings can be united, but they do not unite them.[67]

In other words, geometry and physics do not in themselves unite self and world; they can be treated as languages in themselves without practical application to the world (as has often been the case in science). Weil asks "can I not attain perfect wisdom, wisdom in action, that would reunite the two parts of myself?"[68]

WISDOM IN ACTION

Weil finds it possible to answer the above question positively by emphasizing once again the often neglected features of Descartes' own enterprise. His physics, far from being abstract and disconnected from matter, actually is "packed with matter" and is revealed through the most basic human endeavors. This provides Weil with the means to join the two parts of the human being.

> I can unite them indirectly, since this and nothing else is what action consists of. Not the appearance of action through which the uncontrolled imagination makes me blindly turn the world upside down by means of my anarchic desires, but real action, indirect action, action conforming to geometry, or, to give it its true name, *work. It is through work that reason grasps the world itself and masters the uncontrolled imagination.*[69] (my emphasis)

As long as I allow my imagination to run free, the world is nothing but a mystery that overwhelms me with pleasure and pain. If, however, I use my mind to impose order and sequence on my sensations via the imagination, "if I bring my will to bear only on the idea of a direction, a motion conforming to that act of will immediately follows it; my will leaves its living imprint in the world."[70] The vehicle that imparts this imprint through action is the human body; "in it [the body], and in the world to the extent that I take hold of it by work, the two kinds of imagination are reunited."[71]

In human labor, the human body is used as a tool to turn the world at least partially to the will of the laborer. The human being does not generally have the capacity to produce desired effects in the world directly by bodily effort—the boulder is often too large to be pushed out of the way. Applying geometry to the world, however, can cause one to search for a lever to move the boulder. As Weil observes in *Lectures on Philosophy*, "geometry lies at the basis of a science of work."[72] Through work, geometry in action, the world becomes an obstacle to be overcome, a problem to be solved, not an unknown to be feared.

> It [sensory exploration] turns sensations into signs of distances, sizes, shapes; in other words, signs of work I may do. Conversely, actual work is related to knowledge insofar as it explores the world, not insofar as it changes something in it. . . . To the degree that I can act upon the world through my body and the simplest tools, to that degree I grasp extension itself in my sensations. I am no longer content with drawing geometrical figures; I am practicing geometry.[73]

One of Weil's favorite analogies for how geometry in action through work "grasps" the world is a sailboat tacking against the wind. An obstacle often cannot be overcome through direct confrontation, but an indirect approach guided by geometrical knowledge often is successful. Suppose, for instance, that I need to move the above-mentioned boulder, but it is too large to simply push out of the way. Something indirect must be done—at the outset, I must turn away from the obstacle and apply orderly thinking to the problem. This method requires breaking down the complex problem into its component parts.

> I learn to move only the limb that I want to move, and not the whole body, by a thought. Then, by a kind of geometry in action, I combine these movements according to an order leading from the simple to the complex. . . . As soon as I have taken possession of my body in this way I no longer only conceive, as geometry allowed me to do, that one can tack about in this sea of the world; I do tack about in it. Not only do I have a grasp on the world, but my thought is, as it were, a component part of the world, just as the world, in another way, is part of my thought. From now on I have a share in the universe; I am in the world.[74]

Applying mathematical and geometrical order to the world, then, requires first turning away from the world, just the sailboat can be steered where the pilot wants it to go by tacking against the wind. As Weil writes in a later paper, "In order to think mathematically, we put aside the world; and at the end of this effort of renunciation the world is given us like a bonus."[75]

What sorts of solutions to the problem of the boulder might such mathematical thinking produce? As noted above, each person's primary "tool" for interacting with the world is the body; if my own body is not a sufficient tool to accomplish the task (the boulder is too large), other tools are needed. The application of geometrical and mathematical thinking to the world produces such tools, tools that are in many ways more effective than my own body, since they are tools whose "structure I can change as I like."[76] Tools such as a wheel, winch, and lever transpose the limited power of my body in innumerable ways, all of which become apparent by treating the obstacle as a geometrical problem to be solved. A lever of proper length, intelligently used, can move the boulder, although the action is indirect (pushing on the lever) rather than direct (pushing the boulder). It was this idea of conceiving of the natural world as a system of simple tools that is at the heart of modern mechanics and physics; "science imagines, so to speak, that beneath observed phenomena there are combinations of simple tools, such as those from which machines are made."[77] As Archimedes reportedly said, "give me a point of leverage and I will move the world."[78]

The tools that the worker uses are simple extensions of the human body, which Weil describes as "like a pincer for the mind to grasp and handle the world."[79] But simple tools such as levers are limited in their use. Humanity expands its grasp on the world by combining such simple tools, collectively seeking to bring the complexity of the world more and more into its domain.

> If I want to further extend my domain, I have only one means of doing so, which is to combine simple tools. In this way human action increasingly approaches the infinite complexity of the world, without ever attaining it. Man creates machines with the wheel and the lever, just as he constructs any given point of a conic section with compass and ruler. It is in this way that industry is an extension of work.[80]

Collective work (industry), just as individual work, is the most direct possible contact of the mind and the world; indeed, it is through work that the specter of an irreparable dualistic split between mind and body is ultimately exorcised. This early recognition of the special nature of human work develops and matures throughout Weil's continuing thought. She closes her final work, *The Need for Roots,* with the following concluding thoughts: "All other human activities, command over men, technical planning, art, science, philosophy and so on, are all inferior to physical labor in spiritual significance. It is not difficult to define the place that physical labor should occupy in a well-ordered social life. It should be its spiritual core."[81]

What has this to do with science? Everything, according to Weil, for science and work are essentially the same activity. Just as work is made possible by the human mind mastering the imagination by imposing order on the world via geometry in action, "the purpose of science . . . is to render the human mind master, as far as possible, of the part of the imagination that perception leaves free, and then to give it possession of the world."[82] Science is a collective human attempt, via work and perception, to grasp the world in ways that a single human being cannot. The foundational insight behind Descartes' science is that, by imposing mathematical and geometrical order on our perceptions, it becomes possible for human beings to bring the world that is beyond us within our practical reach.

> Science uses geometrical figures to imitate the perfect connection between what is observed and extension that work establishes. To this end science imagines, so to speak, that beneath observed phenomena there are combinations of simple tools, such as those from which machines are made. Science does not claim that these mechanical models of things duplicate the world. . . . But at least they let us place phenomena that we do not understand in a series with those that we do, ranking them in a geometrical order that proceeds from the simple to the complex . . . and the uniformity of all these mechanical models is determined by what expresses the degree of complexity they have in common, that is, by an algebraic formula.[83]

Science "demystifies" the world, not by bringing it entirely under our control, but by continuing to demonstrate how human beings can increasingly "tack" in the world. "That is why mankind needs science."[84] For instance, although human beings have no power, direct or indirect, over the stars, science brings them into our domain sufficiently that "the pilot dares to use the stars . . . as his instruments."[85] Science understood in this way is no more the exclusive domain of an intellectual elite than is basic human labor; in both activities the human mind is in most direct contact with the world. Furthermore, the dynamic common to work and science is of moral and ethical value. "The application of reasoning to nature, even if it cannot ever be rigorous, is of moral importance and value: through it a man gets the idea of being a worker face to face with nature. Science gives us courage and makes men of us insofar as it is real science."[86]

Weil's vision of human labor in "Science and Perception in Descartes" is idealistic and might be seen as a vision only available to a young woman who had lived a relatively sheltered life to this point and had never experienced real manual work, even though her lifelong affinity for the working classes and the poor began in her early childhood. According to this conception of work, a factory, for instance, would ideally be an environment in which the workers exercised their minds correctly and came to know both the world and their own true natures as thinking and active beings. What she found when she worked as an unskilled worker in a factory for the first time during a leave of absence from teaching for the 1934–35 school year was a situation much farther from this idealistic vision than she could have possibly imagined. As she wrote in a 1935 letter during her factory experience, "What a factory ought to be . . . is a place where one makes a hard and painful, but nevertheless joyful, contact with real life. Not the gloomy place it is where people only obey orders, and have all their humanity broken down, and become degraded lower than the machines."[87]

In "Science and Perception in Descartes," Weil had argued that in order to overcome obstacles presented by the world by means of work, tools are needed to extend the domain of one's body. She described such tools (levers, wheels, etc.) as "human bodies that have no feeling, that I can use anywhere, that are at my disposal, that I can take up, put aside, and take up again, bodies in short that perfectly correspond to the indirect nature of work."[88] Her factory experience revealed to her that in real factories, this is a perfect description of how living, breathing workers themselves are treated.

For these workers, the true factory experience was painful morally and physically, as well as completely dehumanizing.

In many ways, her firsthand experience in the factory changed Weil's life permanently, having a great effect on her future social and political thought as well as introducing concepts such as affliction that play such an important role in her later considerations. Her experience also served to deepen her concerns and insights about science. She continually attributed the dehumanizing effects of technology on human labor to a larger scientific vision of the world that had lost its soul as well as its connection to real human needs and concerns. The Cartesian roots of modern science had led to a science dedicated only to power and manipulation. As we have seen in this chapter, even as a young adult Weil was convinced that the true purpose of science is something quite different than power and manipulation; her continuing investigation of these issues throughout her life was consistently focused on the problem of reconnecting science to its human roots and its true value.

> It is always man's greatness that he recreates his life. He recreates what is given him. . . . By work, he creates his own natural existence. By science, he recreates the universe through symbols. By art, he recreates the alliance between his body and his soul. Note that each of these three creations is a poor, vain, empty thing when taken by itself and unrelated to the other two.[89]

CHAPTER TWO

Classical and Contemporary Science

Classical science . . . seemed to justify the belief that it would be possible, by calculation, measurement, and numerical equivalence, to read throughout all the phenomena occurring in the universe simple variations of energy and entropy, conforming to a simple law. The prospect of such a triumph might well be intoxicating. Soon after came the catastrophe.

—Simone Weil, "Classical Science and After," SN, 10

One of the most important, timeless philosophical issues concerns the relationship between descriptive (what is the case) and prescriptive (what ought to be the case) statements. Among contemporary philosophers, the general consensus is that there is an unbridgeable divide between fact and value; one cannot derive "ought" from "is." Science, whose purview is the realm of what is the case, has nothing to say about values, nor should it attempt to. In a world in which science has become the primary voice of authority, such a split between fact and value creates a crisis for the guidance of human lives. If the predominant voice of authority cannot tell us anything about how our lives should be lived, where is such guidance to be found? What will fill the vacuum?

Throughout her writings, Simone Weil frequently warns that we divide fact sharply from value at our extreme peril. Science cannot simply restrict itself to gathering and analyzing data for the purpose of manipulating future experience; the effort of understanding human experience is value-laden from the start.

> All thought is an effort of interpretation of experience, and experience provides neither model nor rule nor criterion for the interpretation; it provides the data of problems but not a way of solving or even of formulating them. This effort requires, like all other efforts, to be oriented towards something; all human effort is oriented and when man is not going in any direction he remains motionless. He cannot do without values. For all theoretical study the name of value is truth. It is impossible, no doubt, for men of flesh and blood in this world to have any representation of truth which is not defective; but they must have one.[1]

Yet, in Weil's estimation, contemporary science is entirely lacking in orientation toward anything other than itself. Truth has become a lost concept: "present-day scientists have nothing in their minds, however vague, remote, arbitrary, or improbable, which they can turn towards and call it truth."[2] If science is the oracular authority, but no longer considers truth an appropriate or even possible object of pursuit, the notion of truth itself disappears. Since something must fill this vacuum, since human effort must be directed toward something, the purpose of science becomes utility and power, a dangerous prospect.

> Utility becomes something which the intelligence is no longer entitled to define or to judge, but only to serve. From being the arbiter, intelligence becomes the servant, and it gets its orders from the desires. And, further, public opinion then replaces conscience as sovereign mistress of thoughts, because man always submits his thoughts to some higher control, which is superior either in value or else in power. That is where we are today. . . . The tempest raging around us has uprooted all values, disrupted their hierarchy, and put them all in question, to weigh them again on the always untrue balance of force.[3]

Without overestimation, it can be said that Weil's overriding concern in her lifelong thinking about science was to consider how science might be reoriented in such a way as to repair the devastating rupture between fact and value.

In this chapter, we will consider closely Weil's analysis of how contemporary science has come to the unfortunate state described above, primarily through an investigation of several essays collected in *On Science, Necessity, and the Love of God*. As seen in the previous chapter, Weil was convinced in her early adulthood that the roots of classical (post-Renaissance through nineteenth-century) science do not necessitate the development of science as an activity disconnected from human values and practical concerns. How, then, did classical science develop over a period of four centuries into a science in which there is "nothing . . . that a human mind can love"?[4]

Necessity and Entropy

Simone Weil's unfinished essay entitled "Classical Science and After,"[5] written in 1941, is one of the most detailed examples of her mature thinking on the relationship of classical and contemporary science. From its opening pages Weil demonstrates a continuing commitment to understanding science as an activity analogous to human labor, the same commitment revealed in her writings a decade earlier. Her description of the hopes of classical science is reminiscent of her conclusions concerning Cartesian science in "Science and Perception in Descartes."

> Classical science, the science which was revived by the Renaissance and perished around 1900, tried to represent all phenomena occurring in the universe by imagining, between two successive states confirmed in a system by observation, intermediate stages analogous to those traversed by a man executing a simple manual labor. It conceived the universe on the model of the relation between any human action and the necessities which obstruct it by imposing conditions.[6]

The key concept in classical science is *energy*, a concept derived directly from the concept of work. Energy is a function of distance and force, of mass and velocity, providing the common measure of all work. If one represents energy properly in formulae, it becomes possible to establish that the relationship between two experimentally observed states of a system is identical to the relationship between the starting point and completion of human labor. "For all species of phenomena it is sought to establish numerical equivalents between certain measurements made in the course of experiments, on the one hand, and, on the other, the distances and weights which have to be overcome in human labor."[7] As Weil writes in "Reflections on Quantum Theory," the culmination of classical science in the nineteenth century was to apply the concept of energy to all natural phenomena.

> After many experiments . . . nineteenth-century scientists pronounced that in every phenomenon there is increase or decrease in energy which is equivalent to mechanical energy. . . . The fundamental principle of nineteenth-century science is that it should be possible, in the case of every phenomenon, to represent to oneself, at least theoretically, either the production of the phenomenon by means of displacement of a weight or else the displacement of a weight by means of the phenomenon. That is the only meaning of the word energy.[8]

In these later essays, however, Weil's discussion of human labor is much more complex and nuanced than in the earlier texts, no longer containing the idealism apparent in her earlier writings. The human labor upon which classical scientists sought to model the universe is labor of a particular sort.

> The model for this representation is work, or more exactly, the crude elementary form of work, into which practice, knowledge, skill, and inspiration do not enter. In other words, manual labor. . . . The totality of geometrical and mechanical necessities to which the action is always subject constitutes the primal curse which fell upon Adam, which makes the difference between the world and an earthly paradise, the curse of labor.[9]

In what way is the most basic type of labor a curse? Most fundamentally, perhaps, because labor almost always reveals the tenuous relationship between the desire to accomplish a task and the exceptional difficulty in accomplishing that task. Classical scientists had to account for what Weil calls "necessity," a concept, as we shall see, that takes on tremendous importance in all aspects of her later thought. In the present context, necessity refers to the fact that "time has a direction, so that it is never in any circumstances a matter of indifference in which direction a transformation takes place."[10]

This directionality and irreversibility of human action is a fundamental feature of all human experience.

> We experience this necessity not only in the process of growing older, which slowly but unceasingly compels us, but also in the events of every day. It sometimes requires almost no time or effort to knock a book off a table, disarrange some papers, stain one's clothes, crumple some linen, burn a field of wheat, or kill a man; but it takes time and effort to lift a book, put papers in order, clean one's clothes, or launder linen; and it requires a year of labor and trouble to raise a new crop in the field; a dead man cannot be brought back to life, and it takes twenty years to replace a man in the world.[11]

The problem for classical science was to represent this phenomenon, so evident in human experience, in its model of the universe that was to be based on human labor. This aspect of necessity is not represented in the basic elements of classical dynamics; mass, velocity, weight, distance, and energy itself make no reference to the necessity every human act encounters. Furthermore, the physical world itself, apart from human beings, exhibits the same necessity. It takes only a moment for a strong wind to blow a tree over, but it takes years for a new tree to grow. The challenge to classical scientists, then, was to find a way

> to express what is the condition of all human action, to add something to the notion of energy as defined by distance and weight. It has to be added that every transformation has a direction, which is not a matter of indifference. But this has to be expressed in an algebraic formula, in the language of mathematics applied to physics.[12]

The achievement of expressing algebraically the directional necessity of the universe was accomplished through the concept of *entropy*, embodied in the second law of thermodynamics. According to this concept, "it is assumed that every phenomenon involves a transformation of energy such that, once the phenomenon is accomplished, it is impossible by any means or in any conditions to restore the initial state exactly as it was throughout the system."[13] By use of the notion of limit, scientists were able to represent entropy as a function of the increase of energy, volume, pressure, temperature, and mass. After this achievement, the possibility of being able to understand all natural phenomena in terms of variations of energy and entropy according to simple laws appeared to have become an actuality.

In addition to the obvious predictive and explanatory power of the classical model of the universe, Weil draws our attention to what might be called the moral importance of this achievement. The classical model of the universe reveals that, contrary to what we might like to believe, human beings are not the measure of all things.

> Our simplest actions are ruled by a necessity which, when we relate it to all things, presents the idea of a world so totally indifferent to our desires that we feel how very nearly nothing we are. By conceiving ourselves, if one may so express it, from the point of view of the world, we attain to that indifference about ourselves without which there is no deliverance from desire, hope, fear, and becoming, without which one lives in a dream. . . . Classical science . . . is a purification if it is rightly used. It tries to read behind all appearances that inexorable necessity which makes the world a place in which we do not count, a place of work, a place indifferent to desire, to aspirations, and to the good. The sun which it takes for an object of study shines indifferently upon the unjust and the just.[14]

One of the hallmarks of Weil's philosophical thought is her emphasis on the importance of disinterestedness, of escaping from the grasp of what Iris Murdoch calls "the anxious avaricious tentacles of the self."[15] An important lesson to be learned from scientific enquiry is that the world is not at our disposal; rather, we are in many ways at the mercy of the necessity of the world. Recognition of this fact is, for Weil, one of the most important preliminary steps to humility and wisdom.

Yet, Weil continues, there are many problems inherent with the classical model of the universe. One of these problems is that despite its displacement of human beings from the center of all things, this reductionist, mechanistic model of the universe naturally gave rise to hubris of a different sort. The intoxication of having reduced the complex phenomena of the world to a few simple formulae led classical scientists, over time, to believe that "there were no more things in heaven and earth than in their laboratory—and indeed in their laboratory only at the moment when an experiment succeeded."[16] This contributes to precisely the separation of scientific explanation from lived experience that so concerned Weil in the previous chapter. Only a scientist or philosopher convinced that the mechanical model of the universe entirely explains reality could speak of the complete determinism and predictability of the physical world. Ordinary human experience indicates otherwise.

> It is only a physicist who can speak of "the apparent determinism of the macroscopic scale"; on the scale of our senses there is no appearance of determinism except in the laboratory. Ask a meteorologist or a peasant if they see much determinism in storms or rain; look at the sea, and say if the shapes of the waves appear to reveal a very rigorous necessity![17]

Clearly, Weil argues, the classical model of the universe fails to speak to practical human experience in a coherent manner. In her estimation, this failure arises from a fundamental omission in its representation of human labor and aspirations; it is to her analysis of this failure that we now turn.

A MYSTERIOUS COMPLICITY

According to Weil, all systems of knowledge, including scientific systems, "correspond to a certain relation between man and the conditions of his existence." More fully, "The purpose, the model, and the principle of each of them [scientific models] is the relation between some aspiration of human thought and the effective conditions for realizing it; they reconstruct the whole universe according to this relation and try to decipher it in and through the world of appearances."[18] For instance, the aspiration at the heart of classical science is "the aspiration for power over nature so as to be able to realize any desire."[19] This is why classical science is modeled on the dynamic of human work, since in work human beings seek to bridge the gap between their desires (to accomplish a task) and the achievement of that task (overcoming the obstacles imposed by necessity). Here, however, a problem arises, for phenomena such as human desires and aspirations cannot be modeled in natural phenomena as they exist in human work. "There is no question, of course, of imagining any sort of wills at work behind the phenomena of nature because these would not be analogous to human wills; being unattached to bodies, and supernatural, they would be exempt from the conditions of work."[20] How is it possible to effectively use work as a model for natural phenomena if one of the crucial components of labor, human aspiration and desire, cannot be included in the model?

The answer provided by classical science, Weil argues, is to ignore the desire factor altogether. "In order to establish the analogy between the phenomena of nature and work it is necessary to eliminate one of the terms by which work is defined and without which it cannot be conceived."[21] In other words, classical science requires that we imagine "work with no worker, an obstacle which opposes nothing, conditions which do not condition any undertaking."[22] The result is a scientific model of reality that on one level appears to be intimately connected with the most basic of human activities, labor, yet is entirely unrelated to human concerns because it suppresses one of the two basic elements of labor. The classical model of the universe is a model of work in which desire and aspiration do not exist or, at the very least, are entirely ineffectual. As Weil writes, "the universe it describes is a slave's universe."[23]

As suggested earlier, there is a profoundly important truth contained in this model of the world, because in a very real sense we do live in a universe that is indifferent to our aspirations and desires. We are truly slaves, but we are also something more.

> Man, including slaves, is not wholly a slave . . . he is also, at the same time, something quite different. And the world is indeed that world which sets a distance, laborious to cross, between every desire

and every accomplishment, but it too is at the same time something
quite different. . . . It is true that the matter which constitutes the
world is a tissue of blind necessities, absolutely indifferent to our
desires; it is true, too, in a sense, that they are absolutely indifferent
to spiritual aspirations, indifferent to the good; but also, in another
sense, it is not true. If there has ever been real sanctity in the world,
even if only in one man and only for a single day, then in a sense
sanctity is something of which matter is capable; since nothing
exists except matter and what is inscribed in it. . . . We are ruled by
a double law: an obvious indifference and a mysterious complicity,
as regards the good, on the part of the matter which composes the
world; it is because it reminds us of this double law that the spec-
tacle of beauty pierces the heart.[24]

Although the world presents us inexorably with obstacles to overcome, confronting us with necessity that burdens our very existence, the beauty of this
very same world draws us toward all that is of value and goodness. The tension between contrary, even contradictory, features of the human condition
and of reality in general is of ultimate concern in Weil's thought; here, she
draws our attention to the tension between the fact that on the one hand
we are of no significance and yet on the other hand are in "mysterious complicity" with the very world that is indifferent to us. Science, the investigation
of the world, must have both aspects of this apparent contradiction within
its view.

How do we know that there is value and goodness in the world? Simply
because we have direct experience of it. Goodness, sanctity, and justice are
every bit as real as the crushing burden of necessity, even though classical
science, by reducing all phenomena to an analysis of spatial and temporal
data, addresses only necessity.

Space and time are one and the same necessity, sensed in two ways;
and there is no other necessity. . . . And yet the thinking being feels
himself to be made for something other than time and space; and,
since he cannot banish their presence from his thought, he feels
that at least he is called upon to master them, to dwell in eternity, to
embrace and dominate time, to grasp the whole extended universe

at all its points at once. The necessity of time and space is opposed
to this.[25]

Any science that fails to address all of the features of reality is necessarily
incomplete. As Weil writes in *The Need for Roots,* "If justice is inerasable
from the heart of Man, it must have a reality in this world. It is science, then,
which is mistaken."[26] Because of its deliberate suppression of human desire
and aspiration in its model of the universe, classical science provides no possibility of discovering value and goodness in its investigations:

> Classical science takes as its model for representing the world the
> relation between a desire and the conditions for its fulfillment, but
> it suppresses the first term of the relation. . . . In such a picture of
> the world, the good is altogether absent; it is absent to the point
> where one cannot even find a trace of its absence. . . . Therefore
> classical science is without beauty; it neither touches the heart nor
> contains any wisdom. It is understandable that Keats hated New-
> ton, and that Goethe did not love him either.[27]

Contradiction is part of the human condition and of reality; indeed,
Weil writes in one of her notebooks that "contradiction is what pulls, draws
the soul towards the light."[28] The artificial, reductionistic imposition of unity
on contradictory phenomena requires a continuous denial of that feature of
human experience that most directly turns our attention toward value and
goodness.

> Classical science claimed to resolve the contradictions, or rather the
> correlations of contraries, which are integral to the human condition
> and from which man is forbidden to release himself. Science thought
> to achieve this by suppressing one of the two terms. . . . The hopes
> of classical science were both absurd and impious. They were absurd
> because one cannot reasonably hope that a world in which contrar-
> ies are correlated can be explained by suppressing one of every two
> contradictory terms. . . . They were impious because in this world
> man cannot release himself from contradictions, he can only make
> good use of them.[29]

The crisis of classical science arose around 1900 because new discoveries introduced contradictions that could no longer be suppressed or ignored. Quantum theory and Einstein's theories of relativity required a rejection of several features of the classical scientific model of the universe and a rethinking of several other features. In Weil's estimation, the opportunity to reintroduce the search for value and goodness into scientific enquiry was there for the taking. Instead, "soon after came the catastrophe."[30]

QUANTUM STRANGENESS

Simone Weil's profound concerns about the cornerstones of contemporary science, the theories of relativity and quantum theory, should not be understood as a challenge to the theories on a practical, pragmatic level. On an important level, for scientific theories "the proof is in the pudding." A scientific theory is, at the core, simply an extended hypothesis fashioned to explain experimental data and to predict the results of future experiments. The emergence of new data always presents a challenge to existing hypotheses; when such hypotheses can no longer account for data, either the hypotheses are adjusted or are replaced by new ones. The practical question, in other words, is "does the theory work?"

The answer to that question, in the cases of both relativity and quantum theory, is "yes." As we shall see, the underlying implications of these theories are bizarre, to say the least, but there is no question that the practical applications of contemporary science have been exceptionally successful. For instance, without quantum mechanics, a plethora of familiar things, such as television, lasers, computers, and radio, would be impossible. The transistor, without which modern electronics and computer technology would be impossible, is the result of a purely quantum mechanical phenomenon. When Max Planck, one of the fathers of quantum mechanics, presented his seminal paper on photons in 1900, he said that "experience will prove whether this hypothesis is realized in nature." The subsequent century appears to be a resounding confirmation of the truth of the theory; Weil herself in 1941 wrote that the legitimacy of the quantum hypothesis "is based on the number of calculations, and of experiments based on these calculations, and of technical applications based upon the experiments, that have succeeded thanks to the formula."[31]

Weil's reservations about contemporary science, then, are not rooted in worries about whether its cornerstone theories are "correct" or "work." Her concerns are directed at a much deeper issue, one that raises in a new form her continuing question about what science really is and what its true purposes are. From the outset, scientists were baffled by the peculiar underlying assumptions required in order for theories such as relativity and quantum mechanics to "work." Relativity, for instance, introduces bizarre notions such as curved space, a velocity (light) that is at once infinite and measurable, and time considered as a fourth dimension. Quantum mechanics, which seems so definitive and clear-cut in its practical applications, is actually based on uncertainties, probabilities, and philosophically peculiar ideas. These theories dropped a bomb not only on the world of physics, but on our whole conception of reality itself. Weil's analysis of contemporary science is an extended investigation of what happens when the metaphysical framework of science becomes entirely divorced from the intuitive and commonsensical abilities that frame our human conception of reality.

In Weil's estimation, quantum theory is the more disturbing of the foundational theories of contemporary science; by the time Einstein's relativity theory arose in the 1920s, "there was nothing for the theory of relativity to upset, because the quantum theory had already upset everything round about 1900."[32] Furthermore, "there is nothing new or strange in the notion . . . that motion and rest have meaning only in relation to a given system of reference. This idea is to be found in Descartes and, if it was rejected by Newton, he did not reject it as a patent absurdity."[33] In many ways, Weil is almost dismissive of Einstein's theory, writing that "it is a very simple theory, so long as one does not try to understand it."[34] One of the generating forces behind Einstein's theory was the 1881 Michelson/Morley experiment which showed that the velocity of light is constant in all directions, regardless of the relative movement of the measuring system. Faced with such an absurdity, Einstein developed equations that appeared to eliminate the absurdity. Attempts to translate these equations back into common speech, however, have produced bizarre notions such as curved space and time as a fourth dimension. Still, as heavily as such concepts tax the imagination and common sense, they are tame in comparison with the concepts accompanying quantum theory.

Simply put, quantum mechanics is the study of matter and radiation at an atomic level. While relativity uncovers the secrets of energy, gravity, and space-time, quantum mechanics probes the subatomic world. The notebooks from the last years of Simone Weil's life are liberally sprinkled with meditations and observations on quantum theory and its implications. While these thoughts, as well as her articles on science collected in *On Science, Necessity, and the Love of God,* reveal a deep familiarity with the basic components of this confusing theory, there a few specific features of quantum theory that were of particular concern to her. Some of these features are summarized in the paragraphs that follow before we turn to Weil's analysis and critique.[35]

Waves and Particles

At the macroscopic level we are used to two broad types of phenomena, waves and particles. Briefly put, particles are localized phenomena which transport both mass and energy as they move, while waves are de-localized phenomena (they are spread out in space) which carry energy but no mass as they move. For example, physical objects (such as the book in your hand) are particle-like phenomena, while ripples on a lake are waves. Note that in such waves, there is no net transport of water, hence no net transport of mass. Quantum mechanics is a theory of matter that successfully describes atomic physics by uniting (some might say blurring) the dual concepts of waves and particles.

The wave/particle duality is not new; in the late seventeenth century, Huygens theorized that light was composed of waves, while in the early eighteenth century Isaac Newton claimed that light was made of tiny particles. Over time, it became apparent that neither an exclusively particle theory nor an exclusively wave theory could explain all of the phenomena associated with light. By the end of the nineteenth century, scientists had begun to think of light as both a particle *and* a wave. This strange blurring of the distinction between two entirely different concepts has a foundational place at the heart of quantum mechanics. Entities such as electrons, which we would normally think of as particles, can behave like waves under certain conditions, and entities that we would expect to exhibit primarily wave-like properties, such as light and electromagnetic radiation, sometimes behave like particles.

Key to gaining at least a modicum of comfort with quantum mechanics is temporarily setting aside the usual Western mode of thinking which tells us that the same thing cannot be two distinct things at the same time. In quantum mechanics, it can. One way to get comfortable is to imagine light and matter existing primarily as particles; what is wave-like is the probability of where a certain particle will be. This, of course, introduces probability into the picture in an important way, another of the foundational concepts running throughout quantum mechanics.

Planck and Black Body Radiation

One of the early breakthroughs that led to quantum theory was Planck's solution of the problem of black body radiation; we shall see that Simone Weil focuses more on this event in the history of quantum mechanics than any other. Skipping unnecessary and burdensome details, the problem can be described as follows. All objects emit electromagnetic radiation, consisting of a broad range of wavelengths. These wavelengths are described by a distribution curve where the peak wavelength of a "perfect radiator" known as a black body is given by a specific physical law. At ordinary temperatures, this radiation is entirely in the infrared (non-visible) region of the spectrum, but as the temperature rises to high values, more energy is emitted in the visible wavelength region and the object begins to glow. A bar of steel, when heated to high temperatures, will glow and become first red, then white hot; lava is red hot when it spews from an erupting volcano.

The problem of black body radiation arises because classical physical theory was incapable of predicting the actual shape of the emission spectrum of a glowing object. Classical physics said, furthermore, that radiation should have infinite energy at high frequencies, which is impossible. In 1900, the German physicist Max Planck, after discussing the latest findings on black body radiation with a scientist friend, realized that he could, via a clever mathematical trick, derive an equation that fit the most recent data perfectly. This equation stated that the shape of the observed spectrum of black body radiation could be exactly predicted if the energies emitted or absorbed by oscillating electrons were restricted to integral values of hn, where n is the frequency and h is a constant where $h = 6.5 \times 10^{-27}$ erg sec (now called *Planck's constant*). The import of Planck's work was that radiation is not entirely wave-like, as previously thought, but energy transfer occurs in discrete packets of energy, particular rather than continuous. We might say that with Planck's formula energy is "quantized," setting the stage for the development of modern quantum physics.

The Photoelectric Effect

The physics community reacted with intense skepticism to Planck's new idea and its counterintuitive yet logical conclusion, that light is not continuous, but rather is granular or particular. Planck himself, a reluctant revolutionary, preferred to think of his quantized-energy hypothesis as nothing more than a "neat trick" to explain black body radiation. To carry it further, to extend his equation beyond this specific radiation phenomenon, would be to negate the well-established electromagnetic theory generated in the late nineteenth century by Maxwell. It remained for a much less reluctant revolutionary, Albert Einstein, to carry quantum theory to the next step.

In 1905 Einstein, still an obscure and unknown figure, asked what happens when a particle of light strikes a metal. If light behaves as a particle obeying Planck's theory, it should bounce the electrons out of some atoms in the metal and generate electricity. Einstein found, as expected, that the number of electrons ejected from the metal surface per second depends on the intensity of the light, but also that the kinetic energy of the exiting electrons (determined by measuring the retarding potential needed to stop them) unexpectedly does not depend on the light intensity. It was determined that the photoelectron energy is controlled by the wavelength of the light, not its intensity. Einstein hypothesized that if the kinetic energy of the photoelectrons depends on the wavelength of the light, then so must the energy of the light depend on its wavelength. Furthermore, if Planck was correct in supposing that energy must be exchanged in packets restricted to certain values, then light must be similarly organized into energy packets (photons). Light has the nature of a quantized particle, whose energy is given by the product of Planck's constant and the frequency.

Einstein's publication of his results led to the rapid acceptance of Planck's idea of energy quantization, which had not previously attracted much support from the physics community of the time. Planck won the Nobel Prize in 1918 for his quantum theory, followed in 1921 by Einstein for the photoelectric effect. Today we benefit from the applications of the quantum photoelectric effect. Television, for example, is made possible by this

discovery. Television cameras utilize the photoelectric effect to record a picture on a metal surface. The light enters through the lens of the camera, hits
the metal, and creates certain patterns of electricity, which are then con-
verted into television waves that are beamed into home television sets. Unlike
ordinary camera film, which can be exposed only once, this metal can be
used repeatedly, and hence capture moving images.

The Uncertainty Principle

One of the most important features of classical physics is the importance of precise measurement. At the atomic scale of quantum mechanics,
however, measurement becomes a very tricky process. In 1927, Werner Heisenberg proposed that certain pairs of properties of a particle cannot simultaneously have exact values. More specifically, it is impossible to know both
the velocity and the location of an object simultaneously. In the subatomic
realm, the very act of observing an object changes its position and velocity;
the very act of taking a measurement of an atom's system disturbs the system
so greatly that it alters its state, making it qualitatively different from its state
before the measurement was taken. Why is this?

Suppose, for example, that you wanted to find out where an electron is
and where it is going. An electron is so small that to measure its position in
an atom, photons of light must be bounced off it. However, the light is so
powerful that it bumps the electron out of the atom, changing the electron's
position and location. Furthermore, better and better measuring apparati
would be of no avail. The fact that we cannot know the exact position and
velocity of a single electron is not due to experimental error or observational
technique. Nature does not allow a particle to possess definite values of position and velocity at the same time, a principle now known as *Heisenberg's
uncertainty principle*. Since the very foundation of science is the ability to
measure things accurately, Heisenberg's principle flies completely in the face
of classical physics. The best we can do is establish probabilities concerning
these uncertainties.

These are just some of the features of quantum physics that Simone
Weil discusses in her notebooks and articles. Her thoughts are in many ways
incomplete, due to her premature death as well as the continuing development of quantum physics even to the present day. Despite their incomplete

nature, however, what we do have of her thoughts concerning contemporary science is more than sufficient to indicate the precise areas in which she believed twentieth-century science led humanity into danger. It is to this analysis that we now turn.

CONTINUITY AND DISCONTINUITY

One of the primary features of human existence is the juxtaposition of continuity and discontinuity. We divide the passage of time into years, seasons, days, and hours, but its passage is seamless and continuous. We mark off spatial areas of our world, but space itself is as seamless and continuous as time. Yet numbers, which we use to organize and manage space and time, are discrete and discontinuous. Our organizational schemes in chemistry and biology reflect seemingly discontinuous features of the world, discrete elements with no connecting bridges. Although continuity and discontinuity are contradictories, human experience includes both. As Weil writes, "The human mind cannot make do with number alone or with continuity alone; it oscillates between the two. And there is something in nature which corresponds to each of them, for otherwise man could not exist as he is, with a mind which thinks always in terms both of number and of space."[36] Classical science, with its exclusive focus on space, time, and energy, was an attempt to reduce all natural phenomena to the continuous.

> The whole effort of science, ever since Galileo, has consisted in reducing all phenomena without exception to changes in the relationship between space and time, admitting no variable factors except distance, velocity, and acceleration. Space and time can only be represented as continuous quantities; and energy is precisely the idea through which everything can be reduced to space and time.[37]

In Simone Weil's estimation, the scientific crisis that brought about the end of classical science, sparked by the quantum theory and the theory of relativity, "marked the return of discontinuity"[38] and "introduced into science an acknowledged and asserted contradiction."[39]

It is already clear that Weil actually welcomed the return of the discontinuous to science and was entirely comfortable with contradictions in our understanding of human experience and the world in which such experience takes place. For instance, the wave/particle duality discussed above does not seem to have been particularly worrisome to her. In a 1937 letter to Jean Pasternak, she writes that

> The image of waves and the image of corpuscles are incompatible; and what is there extraordinary about that? It shows the need to elaborate a third image to bring together the analogies represented by the other two. And if that should prove to be impossible, I see nothing to be shocked at in the fact that one has to refer to two incompatible images in order to give an account of a phenomenon.[40]

While attempts to impose unity on conflicting theories are not necessarily misguided, "the attempt to impose unity upon the whole of science . . . was a mad ambition."[41] The unexpected appearance of discontinuity at the quantum level forced scientists to address conflicting features of reality that exemplify the contrariety that is everywhere in human experience. "A fortunate necessity, because no human thought is valid unless it recognizes this relation."[42]

The heart of Weil's critique of quantum theory is not that the theory introduced contradiction into the scientific model of the world by reintroducing discontinuity into the classical model that had focused only on continuity. What disturbed Weil, rather, was the manner in which contemporary scientists incorporated this contradiction into their newly developed models of reality. "What is disastrous is not the rejection of classical science but the way in which it has been rejected."[43] When contradiction is encountered in human experience, one must not attempt to eliminate the contradiction by ignoring one of the contradictories (as classical science did) nor by attempting to impose a premature and misguided unity on the contradictories (as, in Weil's estimation, contemporary scientists did). The proper attitude toward contradictories, according to Weil, is contemplation. "In science, when there is something which doesn't fit it, we should contemplate it instead of trying to get round it."[44] Weil's discussion of quantum theory primarily focuses on contemporary science's failure to do this.

Weil writes that "twentieth-century science is classical science with something taken away. . . . What has been taken away is the analogy between the laws of nature and the conditions of work, that is to say, its very principle;

and it is the quantum hypothesis that has removed it."[45] Even though the classical model was problematic because it suppressed one of the crucial aspects of work, it still maintained a vital connection between its model of the universe and human experience. Weil begins tracing how this connection was severed by directing our attention back to the concept of energy. All of the factors involved in the concept of energy—distance, time, velocity, acceleration, and mass—are continuous quantities. "Energy is a function of space, and space is continuous; it is continuity *par excellence*; it is the world conceived from the point of view of continuity; it is things, in so far as their juxtaposition encloses the continuous."[46] As discussed earlier, the goal of mature classical science was to explain all phenomena in terms of energy, in other words in terms of the continuous. As it became possible to investigate the world of the atom closely, however, the classical model failed to suffice because entering the world of the atom means entering the world of the discontinuous.

> Confronted with atoms, the task of classical science was difficult. It was necessary to imagine very small indivisible particles, whose motion was subject to the necessities of classical mechanics; and their movements had to be such that they were united by necessities to phenomena observable on a microscopic scale and also united. . . to phenomena observable on a human scale and whose regular variations had hitherto been the sole subject of science. . . . If, instead of that, one imagines the complicated combinations of movements described by the atoms . . . one is obliged . . . to use the notions of chance, probability, average, and approximation.[47]

The assumptions that served classical science so well on the macroscopic scale fail to account for what is occurring on the microscopic scale, presenting contemporary scientists with a choice.

> It becomes necessary either to establish some link between the two physics or else to abandon one or the other of them completely; or at least this should have been obvious, but it is not what happened. It has been possible to establish a link only by assuming that atoms are subject to necessities different from those of classical physics.[48]

The link between macroscopic and atomic physics began to be forged by Planck's solution of the problem of black body radiation, a link that introduces an entirely new, and (according to Weil) incoherent use of the concept of energy.

The problem is that the conception of energy emitted in discrete packets, or quanta, cannot be reconciled on any level with the fact that any possible conception of energy is based on factors that are entirely continuous. When Planck asks us to "consider energy, or rather action, the product of energy and time, as a quantity which varies discontinuously, in successive jumps . . . known as quanta,"[49] he is asking us to do the impossible. "One can think of things as discontinuous, that is to say, as atoms, but even at the price of implicit contradictions one cannot think of space [nor of energy] in this way. . . . To think of space as discontinuous is like thinking of continuity itself as discontinuous."[50] Quantum physics is a theory at whose heart is a concept (discontinuous packets of energy) that is unrepresentable, either on the model of human work or any other conceivable model. As noted earlier, the reemergence of the discontinuous is neither a surprise nor something to be regretted. What is to be regretted is that quantum theory "solved" the problem of the discontinuous by importing a concept (energy) into the realm of the discontinuous that is only conceivable in the realm of the continuous.

> There are so many signs of the return of science to the discontinuous. There is nothing in the least unnatural in this return, which is a phase of the necessary balancing of two correlated ideas. But what is, without exaggeration, contrary to nature is the use of the discontinuous in contemporary physics, when energy, which is purely a function of space, is divided into atoms. With that, what was still called science in 1900, but must now be called classical science, disappeared, for it has been radically deprived of meaning.[51]

Weil's brief description of Planck's procedure is, in her estimation, a primer on what happens when science no longer is concerned with maintaining any connection, even a tenuous one, with models of reality that are recognizable in human experience. As discussed above, Planck's efforts were simply to solve a particular problem, the problem of black body radiation. By manipulating algebraic formulae, he was able to find a relationship, *hn,* that served the required purpose predicting the actual shape of the emission spectrum of a glowing object. Quoting Planck, Weil emphasizes that Planck's only motivation was problem solving, regardless of the incoherence such problem solving might introduce into the larger model of reality. In Planck's formula, one finds "The constant h, which was to become so famous, and which corresponded to an energy multiplied by a time. Such a constant was meaningless in terms of classical science, but 'it was only thanks to it that the fields or intervals indispensable for the calculation of probabilities could be known.'"[52] Perhaps most disastrous, in Weil's interpretation, was that Planck's constant, originally serving only the purpose of solving a specific problem in physics, turned out to be applicable in cases ranging far beyond black body radiation, as Einstein showed.

> It was for the convenience of calculation in regard to a particular case, black body radiation, that he [Planck] introduced discontinuity into energy. His innovation was prodigiously lucky, because it has been accepted that his formulae are valid for all exchanges of energy between atoms and radiations; that is to say, they are valid generally. Thus the word energy has no longer any relation with weights and distances, or mass and velocity; but neither is it related to anything else, because no other definition of energy has been arrived at; it is related to nothing.[53]

In short, "Planck's formula, composed of a constant whose source one cannot imagine and a number which corresponds to a probability, has no relation to any thought."[54] This is but one example of what Niels Bohr, one of the fathers of quantum mechanics, meant when he famously said that "anyone who is not shocked by quantum theory has not understood it." To further investigate the implications of a scientific theory that has no coherent meaning in the traditional sense, Weil turns her attention to the language of physics, algebra, which plays an unexpected role in contemporary science.

> What is the relation that serves it [contemporary science] as its principle and determines its value? The question is difficult to answer, not because there is any obscurity but because it is embarrassing to reply. The philosophic significance, the profound thought at its

center, are like the Emperor's clothes in the story; to say they do not exist is to be branded a fool and ignoramus, so it is more prudent to call them inexpressible. Nevertheless, the relation which is the principle of this science is simply the relation between algebraic formulae void of meaning, on the one hand, and technology on the other.[55]

THE LANGUAGE OF ALGEBRA

In chapter 1, we saw that even in early adulthood Simone Weil was concerned about how algebra, the mathematical language of physics, can be used in such a way that it becomes a language entirely divorced from what it is intended to represent. In an early letter to her mentor, Alain, she blames Descartes for opening the door to the misuse of algebra.

It appears to me that one might, if one wished, sum up the whole development of the last three centuries by saying that Descartes' venture has turned out badly. . . . Descartes never found a way to prevent order from becoming, as soon as it is conceived, a thing instead of an idea. Order becomes a thing, it seems to me, as soon as one treats a series as a reality distinct from the terms which compose it, by expressing it with a symbol; now algebra is just that. . . . It is only analogy that makes it possible for thought to be at the same time absolutely pure and absolutely concrete. Thought is only about particular objects; reasoning is only about the universal. Through the trick by which it has tried to resolve this contradiction, modern science has lost its soul; this trick consists in reasoning only about conventional symbols.[56]

Weil's analysis of the use of algebra in contemporary science brings us to the heart of what happens when science becomes entirely divorced from any possible representation in human experience or imagination; "What makes the abyss between twentieth-century science and that of previous centuries is the different role of algebra."[57]

In relation to physics, the original purpose and use of algebra is an entirely useful and practical one.

In physics algebra was at first simply a process for summarizing the relations, established by reasoning based on experiment, between the ideas of physics; an extremely convenient process for the numerical calculations necessary for their verification and application.[58]

In this sense, the value of algebra is established on account of its effectiveness in representing in a manageable way the relationships between the objects studied experimentally by physicists. Algebra is problematic as a language, however, because "ordinary language and algebraic language are not subject to the same logical requirements . . . incompatible assertions may have equational equivalents which are by no means incompatible."[59]

Einstein's solution of the problematic results arising from the Michelson/Morley experiment provides an example. It makes no sense in ordinary language to say that the velocity of light is constant in all directions, since a finite speed cannot be constant in all directions if measured from a system which is itself in movement in a certain direction. "Nevertheless, Einstein translated these conclusions, irreconcilable with one another, into algebraic formulae and then combined these formulae, as if they could all be simultaneously true, and derived equations from them."[60] Algebra once played an auxiliary role to common language, but in contemporary science has assumed the primary role. Through manipulation of algebraic formulae, "results may be obtained which, when retranslated into spoken language, are a violent contradiction of common sense."[61]

Once algebra has ceased to be a language that represents anything other than itself, there is virtually no limit to the bizarre and nonsensical notions that can be made sense of in its formulae. Planck's formula which presents us with discrete packets of energy is just one example. More and more the tendency in contemporary science and mathematics is not even to attempt to provide commonsensical models of anything.

Then, one comes to questions where one no longer understands what the difficulty is (transcendent numbers: π, e). One has to take refuge in algebra, and then one has reached the stage when signs have become a fetish: "Imaginary numbers." One doesn't understand why imaginary numbers make calculations easier, and mathematicians do not worry about it. In our days, mathematical invention

is a matter of making calculations easier. . . . Today people have
allowed themselves to become carried away by symbols; they are
arrived at the point of getting results they do not understand.[62]

When the results of contemporary scientific research can no longer be pictured or imagined in any meaningful way, the only justification for science
becomes its technical applications, how well it "works." Any further possible
meaning for science is abandoned. "Once such a state of affairs is admitted,
physics becomes a collection of signs and numbers combined in formulae, which are controlled by their application."[63] The story is told that the
great Hungarian mathematician John von Newmann, when asked by a
young colleague about a particularly difficult mathematical problem, replied,
"Young man, in mathematics you don't understand things, you just get used
to them."[64]

Weil uses several analogies to help us focus more sharply on the real
dangers of algebra as used by contemporary scientists; one of the most striking is in one of her early notebooks, where she writes the following: "MONEY,
MECHANISM, ALGEBRA: the three monstrosities of contemporary civilization.
Complete analogy."[65] In her estimation, all three of the above are monstrosities because they act as levelers, reducing to one plane factors and variables
that exist on different, hierarchical planes. Money, for instance, represents
everything on the same quantitative value plane, entirely failing to commu-
nicate the qualitative values that might not be reducible to cash value and
dollar signs. Mechanism strips human beings of their individual desires,
hopes, and talents, reducing them to mere cogs in a machine. Similarly, algebra reduces to the same level features of reality that very well cannot be unified in any commonsensical or reasonable manner.

If a profound thought is inexpressible it is because it embraces at
the same time several vertically super-imposed relations, and ordinary language is ill-adapted for reflecting differences of level; but
algebra is even worse adapted because it puts everything on the
same level. Proofs, verifications, hypotheses, almost arbitrary con-
jectures, approximations, conceptions of appropriateness, convenience, and probability—all these things, once translated into letters,

play an equal role in equations. If the algebra of physicists gives the impression of profundity it is because it is entirely flat; the third dimension of thought is missing.[66]

The often misguided drive toward unity in science, what some contemporary scientists have called the search for a "theory of everything," is facilitated by algebra, which provides the appearance of unification at the expense of entirely losing sight of the complexities, contradictions, and hierarchies of value and meaning that are at the heart of the human condition. As Weil writes in her Marseilles notebooks, "Science (like every human affair) is situated on several vertical planes. Algebra puts everything on the same plane. The practice of algebra, as of everything else, is limited, and is rendered useless by overstepping the limit."[67]

THE SCIENTIFIC VILLAGE

One might suppose that the reaction of the public toward a science whose results are a direct and consistent violation of common sense would be a strongly negative one. Weil observes, however, in the following extended passage from "Wave Mechanics," that the opposite is the case:

Public attention has been captured by wave mechanics . . . not because of its scientific value, which the public hardly tries to appreciate, but because of its revolutionary character and the way it seems to upset established ideas. Respectable scientists . . . accept wave mechanics because it confers coherence and unity upon the experimental findings of contemporary science, and in spite of the astonishing changes it implies in connection with ideas of causality, time, and space, but it is because of these changes that it wins favor with the public. The great popular success of Einstein was the same thing. The public drinks in and swallows eagerly everything that tends to dispossess the intelligence in favor of some technique; it can hardly wait to abdicate from intelligence and reason and from everything that makes man responsible for his destiny. Which does not mean, of course, that mathematical technique, upon which wave mechanics, like relativity, is founded, can dispense with intelligence; nor can any other technique; but it is such a singular kind of intelligence, so strictly bound to the technique itself, that it seems to have no connection with that other intelligence, common to all, which had so long seemed to be the only judge of fundamental ideas. On seeing it apparently lose that privilege . . . they [the public] ask nothing better, it would seem, than to leave their destiny, their life, and all their thoughts in the hands of a few men with a gift for the exclusive manipulation of this or that technique.[68]

The apparent profundity of theories such as quantum theory and relativity arises from the contradictions and strangeness inherent in all attempts to explain the theories in understandable terms. Simply put, it is supposed that anything difficult to explain or "put into words" must be exceptionally profound and accessible only to a special few. The "profundity" of contemporary science, however, has no more substance than the emperor's new clothes in Hans Christian Anderson's tale.

Philosophers, with respectful enthusiasm, are at pains to interpret what they cannot understand and to translate the equations into philosophy; and commentators in general, both lay and sometimes also even scientific ones, continue with touching perseverence to seek the profound meaning, the conception of the world, enshrined in contemporary science. Quite in vain, because there is none.[69]

Just as no one other than an unsophisticated and uneducated child had the courage in Anderson's tale to simply observe that the emperor had no clothes, so in our contemporary world no one is willing to admit that they find no deep, profound meaning in contemporary science.

Anyone with any pretense to culture would fear to be thought a fool if he acknowledged to others or to himself that he could not perceive the slightest philosophical significance in the innovations of contemporary science; he prefers to invent one, which is inevitably very hazy.[70]

In earlier ages the common person placed his or her destiny into the hands of a religious elite; in the contemporary world the public is more than willing to take a perverse delight in science's violation of reason and to once again place its destiny in the hands of an elite.

Weil argues, however, that this abdication of responsibility and reason is particularly dangerous in our contemporary world. When scientific enquiry is entirely disconnected from human values and derives its only meaning from better and more efficient technical applications, human existence itself is placed at risk. Furthermore, those elite in whose hands human destiny has been placed are entirely incapable of bearing the burden.

> The development of technology offers no hope of happiness until we have learnt how to prevent men from using technology to dominate their fellows instead of nature; and, as for our knowledge, scientific progress can add nothing to it, since it is recognized today that the layman cannot understand anything about science and that scientists themselves are laymen outside their own special departments.[71]

Weil uses passages from Planck's own attempts to draw philosophical implications from his scientific work to illustrate that, in reality, there is no coherent scientific voice of authority. No one is in charge in the "scientific village." Planck writes,

> Contrary to what is readily maintained in certain circles of physicists, it is not true that, in order to elaborate on hypotheses, one may utilize only the ideas whose sense may be defined by measurements, independently of all theory. . . . On the contrary, a measurement receives its physical sense only by virtue of an interpretation made of the theory . . . the inventor of an hypothesis has unlimited scope in the choice of whatever means he may deem helpful to his ultimate purpose. . . . He even invents his own geometry as he goes along. . . . And therefore an hypothesis can never be declared true or false in the light of such measurements. All that can be asked about it is how far it reaches or falls short of serving some practical purpose or other. An important scientific innovation rarely makes its way by

gradually winning over and converting its opponents. . . . What does happen is that its opponents gradually die out and that the growing generation is familiarized with the idea from the beginning.[72]

Weil describes in *The Need for Roots* how Planck's own theory demonstrates how far science has strayed from the notion of truth.

> If people hadn't taken such a great fancy to the quantum theory when it was first launched by Planck, and that in spite of the fact that it was absurd—or perhaps because it happened to be so, for everyone was tired of reason—nobody would ever have known that it was a fertile one. At the time it became so popular, there was absolutely no solid data for giving one any reason to suppose that it would be. Hence we find a Darwinian process operating in science. Theories spring up as it were at random, and there is survival of the fittest. Such a science as this can well be a form of *elan vital,* but certainly not a form of the search for truth.
>
> The general public itself cannot be, nor is it, ignorant of the fact that, science, like every other product of collective opinion, is subject to fashion. . . . We should regard this as a scandal, if we were not too brutish to be sensitive to any scandal at all. How can one possibly accord a religious respect to something subjected to fashion?[73]

Given this, science cannot speak with any sort of authority that transcends taste, style, convenience, and the general contingency of human experience.

The public continues to insist on viewing science as "a sort of supernatural oracle and source of pronouncements which, although certainly differing from year to year, are necessarily each wiser than the last"; scientists themselves "are naturally the first to pass off their own opinions as if they were the deliverances of an oracle, for which they have no responsibility and cannot be called to account."[74] In reality, however, "there is no oracle, but only the opinions of scientists, who are men . . . the state of science at a given moment is nothing else but this; it is the average opinion of the village of scientists."[75] The village of scientists is a closed village, requiring an almost incestuous commitment to the village itself and rejection of anyone's opinion from outside the village. Each villager has a narrow specialization, with little or no ability to raise his or her gaze from that narrow focus.

> The villagers seldom leave the village; many scientists have limited and poorly cultivated minds apart from their speciality or, if a scientist is interested in something outside his scientific work, it is very unusual for him to relate that interest, in his mind, with his interest in science. . . . No one has ever been particularly concerned to develop their critical spirit. At no point in their lives are they specially trained to put the pure love of truth above other motives; there is no organized process of elimination which makes a tendency in this sense a condition of entry to the village.[76]

If Weil's analysis to this point is accurate, it is not surprising that there is no commitment to or love of truth in the scientific village. Contemporary science has no concept of truth, and "as soon as truth disappears, utility at once takes its place."[77]

Writing in the midst of World War II, Weil was well placed to observe directly the devastating effects of utility, technology, and power unrelated to human values.

> Utility becomes something which the intelligence is no longer entitled to define or to judge, but only to serve. From being the arbiter, intelligence becomes the servant, and it gets its orders from the desires. And, further, public opinion then replaces conscience as sovereign mistress of thoughts, because man always submits his thoughts to some higher control, which is superior either in value or else in power. That is where we are today.[78]

Weil was convinced that the crisis described in her analysis of contemporary science is one in which "the entire world is at stake."[79] In a world devoid of values, a world in which technology and power are the only deities, what hope is there? The final lines of "Reflections on Quantum Theory" identify the only possible path back to hope and value.

For a long time now, in every sphere without exception, the official guardians of spiritual values have allowed them to decay. . . . A sort of fear prevents us from recognizing this fact, as though there was a risk, by doing so, of damaging the values themselves; but, so far from that, in the period of sorrow and humiliation which we have already entered and which will perhaps be a very long one, our only hope of recovering some day what we lack is to feel with our whole soul how well-merited our misfortune is. We see intelligence more and more enslaved by the power of arms and everyone today is compelled by suffering to be aware of that servitude; but even before there was any power to obey, the intelligence had already sunk into a servile condition. When someone exposes himself as a slave in the market place, what wonder if he finds a master?

The tempest raging around us has uprooted all values, disrupted their hierarchy, and put them all in question, to weigh them again on the always untrue balance of force. Then at least let us, too, in these days, put them all in question for ourselves, each on his own account; let us weigh them in our own hearts in the silence of true attention and in the hope that it will be granted to us to make of our conscience a true balance.[80]

A new, resurrected vision of science must play an integral part in this process. As Weil wrote in her notebooks, "science, today, will either have to seek a source of inspiration higher than itself or perish."[81]

CHAPTER THREE

Monochords and Bridges

Their sole aim was to conceive more and more clearly an identity of structure between the human mind and the universe. Purity of soul was their one concern; to "imitate God" was the secret of it; the imitation of God was assisted by the study of mathematics, in so far as one conceived the universe to be subject to mathematical laws, which made the geometer an imitator of the supreme lawgiver.

—Simone Weil, *SL,* 117

When she wrote "Science and Perception in Descartes" as a young woman, Simone Weil was not entirely sure about the motivations behind Greek science. Although, as discussed in chapter 1, she credited Thales with revealing for the first time that careful examination of the natural world could lead any human being to practical knowledge, she also recognized that for the Greeks science was merely a "more careful kind of perception."[1] It was something more akin to a metaphysical system and religious practice, rooted in a complex and puzzling attraction to geometrical and numerical relationships. As Weil's thinking about mathematics and science grew and matured,

she came to believe more and more that the Greeks, specifically the Pythagoreans, "as if drunk on geometry,"[2] possessed insights specifically capable of
addressing the shortcomings not only of contemporary science but of our
entire relativistic, power-oriented contemporary worldview. It is to these insights that we turn in this chapter.

It might be said that if the ancient Greeks were "drunk on geometry,"
Weil herself was "drunk on the Greeks." Her obsession with the Greeks is
clearly exemplified by her intense attraction to their science and mathematics, as she herself admitted in *The Need for Roots*. "It will occur to few
people, unless urged on by some particular vocation, to plunge into the atmo-
sphere of Greek science as into something real and vital. But those who have
done so have had no difficulty in recognizing the truth."[3] Weil believed that
Greek science contained in seed form everything that came to fruition in the
classical model of science discussed in the previous chapter. Of more importance, however, was that Greek science contained something that both classical and contemporary science lack, an energy deliberately focused toward
the good and the divine rather than toward technology and power.

> But if Greek science is already classical science it is also at the same
> time something quite different. Plato's famous motto, "None enters
> here unless he is a geometer," is sufficient proof. What people went
> to Plato for was a transformation of the soul, so that it could see and
> love God; who today would dream of employing mathematics for
> such a purpose?[4]

In a 1940 letter to her brother André, Weil asked, "I wonder how many mathematicians today regard mathematics as a method for purifying the soul and
'imitating God?'"[5] The answer to this rhetorical question is certainly "few to
none"; it is difficult for our contemporary minds to even imagine what it
might even mean to regard mathematics in this manner. Yet it is precisely
this, according to Weil, that we must rediscover if we are to return science to
its proper place in the realm of human investigation.

> It was quite different among the Greeks, those fortunate men in
> whom love, art, and science, were three scarcely separate aspects of
> the same movement of the soul towards the good. Compared to

them we are wretched; and yet the thing that made them great is close to our hand.[6]

PYTHAGOREAN PRINCIPLES

The most important philosophical influence on Simone Weil was undoubtedly Plato; hence, it is not surprising that she reveals an almost insatiable interest in the Pythagoreans[7] in her published writings, notebooks, and letters. The Pythagoreans profoundly influenced Plato's philosophy, particularly in his cosmology, myths of the afterlife, the importance of mathematics, and some of the fundamental aspects of his theory of Forms. The Pythagorean influence on Weil's thought is particularly evident in precisely the areas that are of most concern to our present study, in that the Pythagoreans placed mathematics and scientific inquiry within a metaphysical framework that not only provided her with an alternative to the barren metaphysics of contemporary thought, but also provided her with a bridge linking science and mathematics to larger issues unfamiliar to those steeped in the modern and contemporary conceptions of science.

When Weil writes that "both Greek science and classical science are concerned with the same conditions, but the aim of Greek science is totally different; it is the desire to contemplate in sensible phenomena an image of the good,"[8] it is clear that we can expect to find that the entire energy of science as conceived by the Greeks, in particular the Pythagoreans, will be foreign to contemporary ears. The importance of pure mathematics, for the Pythagoreans, influenced disciplines as diverse as physical science, religion, and psychology. In this conception of science, number takes precedence over matter, mathematical accounts of phenomena are preferred to descriptions in terms of physical constituents; hence, the aim of Pythagorean and later Platonic science is different from that of modern science that developed from the roots of empirical, Aristotelian science. Pythagorean science is not so much involved with the investigation of things as with the investigation of the principles underlying the phenomenal world. In addition, the Pythagorean approach to number, as we shall see, elevated mathematics to a subject that must be studied, not for its own sake, but as leading to an understanding of reality. Above all, the Pythagoreans believed that science, religion, art,

politics, and ethics are not at odds with each other; it was this belief in the
unification of disparate elements of investigation and of reality itself that perhaps most of all is responsible for the increasingly Pythagorean nature of
Weil's understanding of mathematics and science.

Weil's writings are filled with fragments from Pythagorean sources;
the following passages from Philolaus, one of the first to record Pythagorean
teachings in writing, and from Plato provide a useful orientation to the Pythagorean mind-set.

> *Philolaus*: Clearly the order of the world and of things contained
> therein has been brought into harmony, starting from that which
> limits and from that which is unlimited.[9]

> The following concerns nature and harmony. What is the eternal
> essence of things, and nature in itself, can be known only by the
> divinity and not by man, or else man may know only this. Not even
> one of the realities could be known to us if there were not the basic
> essence of things, of which the order of the world is composed, some
> limiting, the others unlimited. Since the principles which uphold
> all are not alike, nor of the same root, it would be impossible that
> the order of the world should be based upon them if harmony were
> not brought in by some means.[10] One can see what powerful effect
> nature and the virtue of number has not only in religious and divine
> things but everywhere in human acts and reasonings both in the
> working of various techniques and in music.[11]

> Plato, *Philebus* 16b: The realities called eternal derive from the one
> and the many, and carry, implanted within them, the determinate
> and the indeterminate. We should therefore, since this is the eternal
> order of things, seek to implant a unity in every kind of domain. . . .
> What at the beginning appeared not only as one, and many, and
> unlimited at once, appears also with a definite number.[12]

> *Epinomis* 990d–e: It is the assimilation of numbers which are not by
> their nature similar become manifest in accordance with the des-
> tiny of plane figures; and for any one able to understand, that con-

stitutes a marvel (miracle) not of human but divine agency. . . . But what is divine and marvellous to those who are watchful and whom *the spirit pierces* is how the whole of nature is stamped with kind and species according to each analogy, like power and its opposite continually turning upon the double (i.e. arithmetical, geometrical and harmonic means).[13]

As shown in the above passages, key features of the Pythagorean worldview include number, limited and unlimited, harmony, proportion, and geometry. In order to begin to understand Weil's belief, which she shares with the Pythagoreans, that "eternal mathematics . . . is the stuff of which the order of the world is woven,"[14] we now turn to these features.

Number

Philolaus reportedly claimed that "all that is known involves numbers, for without number nothing can be thought or known,"[15] revealing that the Pythagorean understanding of Number was quite different from the predominately quantitative understanding of today. For the Pythagoreans, Number was a living, qualitative reality which must be approached in an experiential manner. The typical modern understanding of number is that numbers are signs, tokens used to represent quantities or amounts of things. The Pythagoreans, however, believed that Number is not something primarily to be used; rather, the nature of Number is to be discovered. They believed that Number is a universal principle, as real to them as electromagnetism or sound is to us. Number is fundamental to all things: the basic features of all things are numerical, numerical considerations are basic in understanding all things, and all things are generated in a similar way to numbers. According to Philolaus, "number gives a body to things."[16]

In addition, Number included the notion of function for the Pythagoreans. Function is a law of variation and relationship, best illustrated by returning briefly to Thales' measurement of the height of the pyramids, mentioned at the beginning of chapter 1. In Pythagorean mathematics, the *gnomon* is an important feature; originally, the gnomon is the fixed, vertical arm of a sundial, remaining constant through the changes of its shadow over the course of a day. This relationship of fixed gnomon to variant shadow can be mathematically expressed, as Thales showed when he demonstrated that the variant shadows of a human being and the pyramids vary in the same mathematically expressed manner in comparison with the height of their respective gnomons. This union of fixed and variant features via function takes on great importance, both mathematically and metaphysically, as we shall see in the following pages.

Because their science possessed a sacred dimension, the Pythagoreans saw Number not only as a universal principle, but also as a divine principle. This complex attitude toward number is summarized by the following from Weil's notebooks: "To read numbers in the universe and to love the universe: the two things go together."[17] Most fundamentally, as Philolaus claims, "Unity is the principle of everything";[18] as Weil reports, "it is known that among the Pythagoreans one is the symbol of God."[19] For the Pythagoreans, Unity or One is best understood as the principle underlying number rather than a specific number *per se*, as the following from Theon of Smyrna indicates:

> Unity is the principle of all things and the most dominant of all that is: all things emanate from it and it emanates from nothing. . . . Everything that is intelligible and not yet created exists in it; the nature of ideas, God himself, the soul, the beautiful and the good, and every intelligible essence, such as beauty itself, justice itself, equality itself, for we conceive each of these things as being one and as existing in itself.[20]

If all things arise from One, the principle of Unity, then the beginning of multiplicity and separateness begins with Two, the Dyad. This duality is revealed in the duality of subject and object, knower and known. The world is generated out of the relationship between Unity and multiplicity; as Weil describes, for the Pythagoreans "Number is the specific relation of each thing to God, who is unity. The universal ratio is the Logos, the divine Wisdom, the divine Word, with which the universe is in conformity through love."[21] The order of the world, the *kosmos,* is intimately connected with this web of relationship between One and Many. This brings the discussion to the most basic and universal principles of Pythagorean metaphysics, Limit (*peras*) and the Unlimited (*apeiron*).

Peras **and** *Apeiron*

One of the most important features of the Pythagorean account of the universe is its account of the origin of *kosmos,* in which the *kosmos* resembles number, geometrical figures, and the musical intervals by being the product of the imposition of limit on the unlimited.[22] Numbers, geometrical figures, the physical *kosmos,* and musical scales are generated similarly: all come to be when limit is imposed on the unlimited. All are instances of order, perhaps even of sequential order, which exists in different realms.

In many ancient metaphysical paradigms, Matter (the Indefinite, *apeiron*) receives and is shaped by Form (Limit, *peras*). These two principles of *peras* and *apeiron* are hence the most universal and essential elements necessary for the manifestation of phenomenal reality. According to the Pythagoreans, the *kosmos* is compounded of these elements. Aristotle preserved a list of the various ways in which *peras* and *apeiron* are exhibited in the *kosmos* in the "Table of Opposites" included in his *Metaphysics*:

Limit	Unlimited
Odd	Even
One	Plurality
Right	Left
Male	Female
At rest	Moving
Straight	Crooked
Light	Darkness
Good	Bad
Square	Oblong[23]

For the Pythagoreans and Plato, *peras* and *apeiron* are woven together through numerical harmonic relationship; in the *Philebus*, Plato reports that the ancients believed that "the world is a fabric woven of limit and the unlimited."[24] The result of this weaving is the phenomenal universe, in which everything is a mixture of universal constants and local variables.

Since Limit is eternal and fixed, while what is unlimited is evanescent and changing, the limiting aspect of reality is reflective of divinity. Weil summarizes this Pythagorean belief in the following passage from her New York notebook: "Things are natural, temporal; but the *limits* of things come from God. That is what the Pythagoreans say. There is the unlimited and that which limits, and that which limits is God. Consequently limits are eternal."[25] In addition, the interweaving of *peras* and *apeiron* is what gives the *kosmos* its order and beauty, as well as its unpredictability and diversity.

> The Pythagoreans used to say that the universe is constructed out of indeterminateness and the principle that determines, limits, arrests. It is the latter which is always dominant. The tradition concerning the rainbow—surely borrowed by Moses from the Egyptians—expresses in the most touching way the trust which the order of the world should inspire in Man. The rainbow's beautiful semicircle is the testimony that the phenomena of this world, however terrifying they may be, are all subject to a limit. The magnificent poetry of this text is designed to remind God to exercise his function as a limiting principle.[26]

The Mysteries of the Monochord

The Pythagoreans say many cryptic and mysterious things about *peras* and *apeiron* and the manner in which the fabric of the world is woven from the ordered imposition of limit on the unlimited. An illuminating practical illustration of the interaction between *peras* and *apeiron* is provided by one of the greatest of the Pythagorean accomplishments, the discovery of the mathematical proportions and relationships underlying musical harmony. Tradition tells us that Pythagoras discovered the musical intervals, in particular the fact that harmonic, concordant musical intervals can be expressed mathematically. This, as many of the stories concerning Pythagoras, may be a folk tale. There is no doubt, however, that the Pythagoreans investigated acoustic phenomena by experimenting with the monochord, a one-stringed musical instrument with a moveable bridge (figure 3.1).

Figure 3.1

In the monochord, changing the position of the bridge changes the pitch produced by plucking or bowing the string, which remains under the same tension. Since there are an unlimited number of possible positions for the bridge, there are also an unlimited number of possible pitches. In a real sense, then, the monochord string represents an indefinite continuum of tonal flux which may be infinitely divided. Experimenting with possible positions of the bridge, the Pythagoreans discovered that by using Number as the limiting power, nodal points can be located on the string which form the mathematical architecture underlying musical scales and harmonies. Plato summarizes these findings in the following passage from the *Epinomis,* to which I will refer in the following discussion:

> All of nature is marked by the form and the essence of the relationship of opposites, according to each proportion. First that of the numerical double, the relation of one to two, which is double. . . . And in the relationship between one to two there are the means, the arithmetical mean, at equal distance from the least and the greatest, the harmonic mean, which exceeds the least and is surpassed by the greatest according to the same ratio—thus eight and nine between six and twelve . . . between these two means, situated at equal distance from both, is found the ratio of which I speak, by which men have received a share in the knowledge of the agreement of voices and of proportion in the apprenticeship of rhythm and of harmony. This is a gift from the blessed choir of the Muses.[27]

The tone produced by plucking the string is generated by the vibrational frequency of the string; the length of the string itself does not matter, nor does the material the string is made out of, nor does the tension of the string, just as long as it produces a tone when plucked. Let us suppose that the length of the monochord string is *x,* its vibrational frequency when plucked is 6 (for the sake of mathematical ease later on), and the note produced when the whole string is plucked is C (figure 3.2).

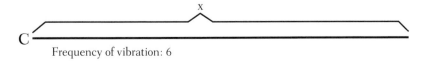

C

Frequency of vibration: 6

Figure 3.2

What happens when the bridge of the monochord is placed exactly halfway between the fixed ends of the string, dividing the string in the proportion of "the numerical double, the relation of one to two"? The note produced is one octave higher than C, which we shall call C'. It is interesting that the vibrational frequency of the string is doubled when the string is halved; this inversely proportional relationship between vibrational frequency and string length holds no matter where the bridge is placed on the string. The ratio, then of the vibration of $1/2x$ (C') to that of x (C), is 12:6, or 2:1 (figure 3.3). The octave C–C' is the basic "field" that needs to be harmonized.

1/2 x

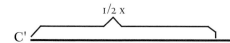

C'

Frequency of vibration: 12
Ratio of C' vibration to C vibration: 12:6, or 2:1
C–C' interval: octave

Figure 3.3

To continue experimentation with the monochord, what happens if the bridge is moved to a point that divides the original string into segments of 2/3 and 1/3 ($2/3x$)? We once again find that the relationship between the dividing of the string and the resulting vibrational frequency of the divided string is inversely proportional: the vibrational frequency of $2/3x$ is 9, which, in relation to 6, the vibrational frequency of x, is 9:6, or 3:2. The tone produced is G; the interval C–G is the *perfect fifth*, the most powerful and fundamental musical relationship. The mysterious mathematical underpinning of harmony is magnified by the fact that 9 is the *arithmetic mean* between 6 and 12; as Plato puts it, the arithmetic mean is "at equal distance from the least and the greatest" ($12 - 9 = 9 - 6$). More generally, the arithmetic mean between any two numbers A and B is found by $(A + B)/2$. In other words, the vibrational frequency of the tone producing a perfect fifth "just happens" to be the arithmetic mean between the vibrational frequencies of the two end points of the octave (figure 3.4). The perfect fifth is, in a very real sense, the arithmetic mediation placing limit on the undifferentiated number of tones between C and C'.

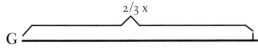

G

Frequency of vibration: 9
Ratio of G vibration to C vibration: 9:6, or 3:2
C–G interval: perfect fifth
9 is the arithmetic mean between 6 and 12: (6 + 12)/2 = 9

Figure 3.4

What of the *harmonic mean,* which is found between any two numbers A and B by the equation $2AB/(A + B)$? In our octave C–C', with endpoint frequencies 6 (A) and 12 (B), the harmonic mean produced by the above formula is 8, satisfying Plato's description of the harmonic mean as the mean that "exceeds the least and is surpassed by the greatest according to the same ratio" [(8 – 6)/6 = (12 – 8)/12)]. This vibrational frequency is, in relation to 6, the vibrational frequency of x, 8:6, or 4:3. In keeping with the above findings that the length of the divided string and the resulting frequency are inversely proportional, we should expect that vibrational frequency 8 will be produced by plucking string $3/4x$, and this indeed turns out to be the case. The tone produced by $3/4x$, with vibrational frequency 8, is F; the interval C–F is the *perfect fourth,* the next most powerful musical relationship after the perfect fifth (figure 3.5).

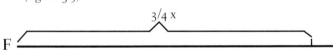

F

Frequency of vibration: 8
Ratio of F vibration to C vibration: 8:6, or 4:3
C–F interval: perfect fourth
8 is the harmonic mean between 6 and 12: [2(6)(12)]/(6 + 12) = 8

Figure 3.5

Finally, comparing the vibrational frequencies of the four tones identified thus far reveals even more basic mathematical relationships (figure 3.6). The four tones we have generated, arranged in the mathematical relationship shown in figure 3.6, provide enough information to complete the musical scale. Notice that not only is 6:9 (C–G) a perfect fifth, but so is 8:12 (F–C'), since 6/9 = 8/12. In addition, not only is 6:8 (C–F) a perfect fourth, but so is 9:12 (G–C'), since 6/8 = 9/12.

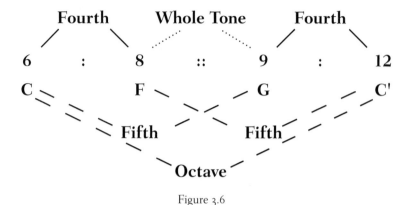

Figure 3.6

The third important mathematical mean, the *geometric* mean, is also revealed in the above diagram. The geometric mean between any two numbers *A* and *B* is found by the equation $\sqrt{(AB)}$, which in this case is $\sqrt{(6)(12)}$, or $\sqrt{72}$. Although this does not specify a whole number, there is an important symmetry present, for the geometric mean between 6 and 12 ($\sqrt{72}$) is the same as the geometric mean between the harmonic mean (8) and arithmetic mean (9), since $\sqrt{(8)(9)}$ is also $\sqrt{72}$. As Weil writes, "the scale does not contain the geometric mean as a note, but is symmetrically disposed around that mean; there is the same geometric mean between one note and its octave and between the fourth and the fifth."[28] The importance of this additional symmetry is magnified when it is discovered that the ratio 8:9 (*F–G*) in the middle of the octave defines the *whole tone*. It is this proportion that Plato refers to as "the ratio . . . by which men have received a share in the knowledge of the agreement of voices and of proportion in the apprenticeship of rhythm and of harmony." Once this proportion has been defined, it is a matter of little difficulty to fill in the rest of the notes of the scale.

These examples illustrate the Pythagorean discovery that music results from the imposition, by means of number, of order and limit on the unlimited continuum of possible tones. It is difficult to imagine how wonderful and surprising it must have been to learn that the fundamental features of music could be expressed numerically. The experiments conducted by the Pythagoreans on the monochord confirmed the importance of numerical *peras* as the limiting factor in the otherwise indefinite realm of manifestation of musical tones. It also suggested for the first time that if a mathematical harmony underlies the realm of tone and music, then Number may account for other phenomena in the cosmic order, from psychological principles to planetary motion.

Harmonia

The Pythagorean use of the term *harmonia* provides further insight into the vast sweep of their mathematical metaphysics. *Harmonia* meant first a fitting together, a connection or joint; later it meant the tuning of a lyre or even the string that is being tuned. Essentially, *harmonia* refers to the ways in which the *kosmos* is ordered rightly, the ways in and the points at which the imposition of limit on the unlimited gives rise to an ordered whole. This principle identifies the fitting together of the various pairs in Aristotle's "Table of Opposites" throughout the ordered *kosmos*.

Crucial to the idea of *harmonia* is the notion of mediation.[29] As exhibited with the examples of arithmetic and harmonic means in the previous section, the middle (mean) term in a mathematical proportion links two extreme terms together—this precise linking together, declaring an identity of ratios, is *harmonia*. There is only one appropriate arithmetic or harmonic mean between specific extremes, a precision exemplified in the harmonies of music. The precision and aptness of *harmonia* is described well, with a particular example, in the following passage from Weil:

> The Pythagoreans said that harmony or proportion is the unity of contraries, *qua* contraries. There is no harmony if the contraries are brought together forcibly, nor if they are mixed; the point of unity has to be found. Without ever doing violence to one's soul, without ever seeking either consolation or suffering, to contemplate the thing, whatever it is, that rouses emotion, until one arrives at the secret point where sorrow and joy reach the point of purity where they are one and the same thing; that is the essential virtue of poetry.[30]

The central focus of Pythagorean thought is in many respects placed on the principle of *harmonia*. The *kosmos* is One, but the phenomenal realm is a differentiated image of this unity—the world is a unity in multiplicity. What maintains the unity of the whole, even though it consists of many parts, is the hierarchical principle of *harmonia,* the principle of relation, which enables every part to have its place in the fabric of the whole. François Lassere summarizes the Pythagorean understanding of *harmonia* as follows:

> It drew its strength from the profound consciousness of an all-embracing harmony, by which life, movement, existence and the cosmic spaces themselves were all alike made subject to the law of Number alone. . . . Their conviction was in the first place a matter of feeling rather than reason . . . less a mathematical proof than a hymn of praise sung in honor of the divine architect and the perfection of his handiwork.[31]

We shall see that it is the complex notion of *harmonia* that perhaps attracted Weil the most to the Pythagorean metaphysic and caused her to carry this principle into realms that the Pythagoreans themselves might not have considered. Her constant insistence on the connectedness and ultimate unity of all phenomena, an insistence that places science in union with religion, art, philosophy, and social interaction, is fundamentally Pythagorean. And, as we have seen, *harmonia* is exhibited both abstractly and pragmatically. The harmonic proportion exists as a purely universal principle, but it would have never been discovered without empirical experimentation on the monochord. The value of the harmonic proportion lies in *both* its universal nature as well as the significance and usefulness of its particular applications. In Pythagorean thought, and for Weil, through the creative dialectic between the temporal and the eternal, there necessarily occurs a form of integration between otherwise purely theoretic and pragmatic approaches.

REALITY CHECK

Before proceeding to an investigation of how these Pythagorean principles influence and are apparent in Simone Weil's own attitudes toward mathematics, science, and metaphysics, a "reality check" is in order. It is, to say the least, a challenge to the contemporary mind to take the Pythagorean beliefs concerning these issues seriously. Mathematics reflective of divinity, a *kosmos* consisting of materials arranged in harmonic proportions by a divine geometer, offering sacrifices in gratitude for discoveries in geometry—these are features of and activities in keeping with a worldview that, at best, provided a meaningful starting point for a science that, over many centuries, developed into the modern science of the sixteenth through the nineteenth centuries and the contemporary science of the twentieth century and beyond. At the same time, to take some of the more unusual aspects of Pythagorean thought seriously seems tantamount to continuing to worship Zeus and the pantheon of Olympic deities. As Richard McKirahan writes, "their [the Pythagoreans'] failure to distinguish between the nature of numbers and the nature of material objects . . . leaves them open to charges that their cosmogony attempts the impossible, to make numbers the physical constituents of material things."[32] In short, "the Pythagoreans literally did not know what they were doing."[33]

What are we to make, then, of Simone Weil, an important twentieth-century thinker who, as we have already seen and will continue to see, not only takes many of the basic features of Pythagorean and Platonic science and mathematics seriously, but also considers them to be required pieces of the foundation for a new understanding of science in our age? Rush Rhees, one of the few readers of Weil who has had much to say about her philosophy of science, asks precisely this question. In *Discussions of Simone Weil*,[34] Rhees brings Wittgenstein into dialogue with Weil concerning many of the issues germane to our present study. A brief consideration of this interaction will serve to critically focus our future discussion, for Weil's attempt to bring the insights of the Greeks to bear on a reconstruction of contemporary scientific paradigms will be an utter failure if it cannot withstand the challenge of an astute thinker such as Rhees.[35] Despite his being steeped in the analytic philosophical tradition, Rhees finds himself strangely attracted to aspects of Weil's thought that, from his perspective, elude meaning and interpretation altogether.

Rhees' critique is of particular importance, because he is not at all inclined to dismiss Weil as a "crank" of sorts, even though her writings appear to leave her wide open to any number of straightforward philosophical criticisms. He does not dismiss such writings, because he believes that the failure may well be with him rather than with her. "Her remarks . . . often invite academic criticism and seem wide open to it. But when I do try to criticize them in this way, I always suspect that I am being stupid, that I am failing (unable) to read what she has written in the way that it should be read."[36] Rhees wants to give Weil the greatest benefit of a doubt, because he cannot dismiss the insights of anyone as capable of brilliance as Weil obviously is.

> On a very great number of questions I have found remarks of hers which are more illuminating than any others I know. . . . She is an example of a writer whose greatness of soul comes out in frequent remarks, which makes the difficulties or shortcomings of her more theoretical observations seem of no importance at all. Remarks in which she shows a soul or character before which I can only stand still, which I could no more imagine than I could have imagined Bruckner's music if I had not heard it. Something like gratitude (it would be stupid to speak of admiration).[37]

It is difficult, however, to give unlimited acceptance to someone who regularly equivocates in her use of any number of important concepts and terms.

> Weil thinks, apparently, that questions expressed in religious meditation or in remarks concerning moral judgments or difficulties have some important connections with questions in the philosophy of science, questions regarding the application of mathematics and of physical theories, the relations between ideal constructions and the natural processes or obstacles we run against.[38]

Indeed, it is Weil's frequent failure to distinguish between obviously different uses of terms and concepts that creates the greatest difficulty in interpretation for Rhees.[39]

Consider Rhees' discussion of Weil's treatment of "beauty," for instance.

> We have to remember that [according to Weil] the more complete or true methods of science would be designed to show the beauty of the world, as, Weil would contend, Greek science was. It is not just that she had not worked out her account of Greek science as she had intended to do. . . . She speaks of the "beauty" of mathematical

proofs, of the beauty of a natural science, of the beauty of a dramatic tragedy, of the beauty of a piece of music—as though anyone could see that one means the same in each case.[40]

Yet obviously, Rhees contends, "beauty" means entirely different things in these different contexts. "I do not stand and contemplate a mathematical proof as I do a landscape. Nor listen to it as to a piece of music or a drama."[41] For that matter, can one even meaningfully use the term "beautiful" in describing a mathematical proof?

> "Nobody has ever tried to *sing* the calculus." Beethoven wrote on his *Missa Solemnis:* "It comes from the heart; may it return to the heart." Can one say this about a mathematical proof? Certainly one can find an aesthetic delight in the symmetry, the neatness or economy and the elegance of a mathematical proof. An impression of *perfection.* Everything falls into place and nothing is wasted. There is nothing *cumbersome* about it. And so forth. But *beautiful?* Something which comes from the heart or goes to one's heart?[42]

Rhees knows, of course, that Weil's manner of speaking about mathematics and science is heavily indebted to the Pythagoreans. This is unhelpful to him, however, because he finds their perspective as alien to his own as is Weil's.

> She adopts a Pythagorean view of mathematics, according to which pure mathematics is simultaneously a formal calculus and discipline, a theory of nature and of natural happenings, and a religious mystical doctrine. But I have *no* idea what it means, and I have just to back away and sit down. I do not know *what* Pythagoras recognized when he celebrated his geometrical discovery about triangles inscribed in a semi-circle with a religious feast.[43]

Hence, Rhees is frequently reduced to allowing simply that "she [Weil] sees in geometry what most of us do not and cannot see there, especially in the idea of arithmetic proportion . . . although it may be that Pythagoras did."[44]

Rhees perhaps would like to dismiss Weil as a sloppy thinker,[45] refraining from doing so only because he suspects that there is another plane of meaning on which her comments would make perfect sense, a plane that unfortunately is unavailable to him as it is perhaps, in his estimation, outside the realm of ordinary non-religious discourse.

> She [Weil] would tell me this shows how little I understand, and I am sure she would be right. And no doubt I should leave it there: I do not understand. Full stop. But can someone tell me the address of someone who *does* understand so that I might ask him *how* I ought to look at her writings?[46]

Rhees comes to at least a tenuous peace with Weil's apparent equivocations by doing what any good Wittgensteinian would do in the presence of what "cannot be said"; he decides that Weil is working and thinking on a level accessible not by language, but by something outside the realm of ordinary discourse altogether.

> Much of what Weil says about necessity, about the use of mathematics in the study of science, the study of what things are, how they are related to one another, much of this is an expression of something which could not be understood except by someone who had known the grace of God as she did. It needs not only religious faith, but a kind of religious insight, in order to understand the phrases or the figures or the grammar of what she writes.[47]

Rhees ultimately deals with his failure to comprehend what Weil is saying about science and mathematics by raising her to a plane of existence available to only a select few, while at the same time placing her thinking on these issues conveniently outside the barrier of philosophical discourse and criticism.

> However much I study her later writings on *science,* I do not think I have learned anything at all. I am inclined to say that what she wrote then is not philosophy but religious meditation. This would not be anything against it, and my remark would not be meant as derogatory. . . . But her greatness lay in her meditations on moral and religious questions. And I doubt if 'greatness' is the word to use

here. I think of her as a saint. This shows in her life, as far as I know of it. . . . And it is true of her essays on science. It is this which makes it particularly hard to know how to read or criticize various things she says.[48]

From Rhees' perspective, ordinary discourse requires a careful distinguishing between the uses of words, concepts, and ways of speaking; the possibility of communication depends on this. Upon encountering an obviously powerful and profound thinker who regularly violates these rules of discourse, he is faced with the choice of either rejecting that thinker as incoherent or of accepting that thinker as operating at a level of insight available only to those who are, for whatever reasons, operating on a plane outside the realm of ordinary discourse. In Weil's case, he chooses to call that plane "sainthood." At least with reference to this aspect of Weil's thought, he concludes that he must come to a "full stop," he must "sit down," he must "back away." "What we cannot speak about we must pass over in silence."

Is this the only possible conclusion for a contemporary thinker wishing to engage Weil's writings on science and mathematics as fully as possible? Simone Weil herself would certainly not have thought so. Her response to Rhees' difficulties would begin, as does so much of her thought, with identifying the beginning assumptions that support his framework of interpretation. Rhees' (unstated) beginning assumptions are undoubtedly similar to those described by Weil in the following comment concerning Pascal:

> Pascal, when he was on the point of discovering the algebraic form of the integral calculus, abandoned algebra and geometry because he desired contact with God. Today, we cannot imagine that the same man could be a scientist and a mystic except at different times in his life. If a scientist has some leaning towards art or religion it is kept absolutely separate from his main occupation, or if he tries to connect the two he does it, as we know from more than one example, by vague and significantly banal commonplaces.[49]

Why would Pascal, or anyone, believe that algebra and geometry could not facilitate a desire for contact with God? More, why would a person believe that mathematical pursuits are in some way *incompatible* with a desire for contact with God? Such convictions can only arise from more fundamental, metaphysical commitments, commitments that Rhees arguably shares with Pascal. What are these commitments and beliefs?

Rhees' comments concerning Weil's philosophy of science reveal his conviction that distinct disciplines such as science, philosophy, and religion have distinct manners of discourse, perhaps even distinct criteria of truth and meaning. When he reveals his frustration at Weil's "mixing up" philosophy and religious meditation, he also reveals his conviction that "obviously" such seemingly different activities must be kept distinct. When she speaks of mathematics and science in language "obviously" more appropriate to religion or aesthetics, she is violating this conviction even more directly. Rhees would undoubtedly have entirely understood and agreed with Pascal's decision to leave mathematics behind in order to pursue contact with God, not because the one pursuit is necessarily better than the other, but because the two are essentially incompatible pursuits. "We cannot imagine that the same man could be a scientist and a mystic except at different times in his life." It is not surprising, then, that Rhees has difficulty understanding Weil's apparent commitment to not distinguishing between disciplines and forms of discourse that "obviously" must be kept separate.

If Rhees' belief that religion, philosophy, science, and aesthetics cannot be complimentary activities is correct, then Weil's use of Pythagorean and Platonic metaphysical assumptions in her discussions of science is exceptionally problematic. Her conviction appears to be that these various activities are *not* best viewed as distinct and isolated activities. Rather, they are best understood as distinct activities with a unified purpose, a purpose quite different than contemporary thinkers might suppose.

> The whole of Greek civilization is a search for bridges to relate human misery with the divine perfection. Their art, which is incomparable, their poetry, their philosophy, their science (geometry, astronomy, mechanics, physics, biology) which they invented, were nothing but bridges.[50]

Of mathematics in particular, she writes that "one does double harm to mathematics when one regards it only as a rational and abstract speculation.

It is that, but it is also the very science of nature, a science totally concrete, and it is also a mysticism, those three together and inseparably."[51] Any attempt to reconnect science and mathematics with aesthetic and spiritual energies will require an entire rethinking of what science and mathematics are.

> In the indifference which, since the Renaissance, science has shown for the spiritual life, there seems to be something diabolic. It would be vain to try to remedy this by an attempt to maintain science in the realm of nature alone. It is false that science belongs wholly to that domain. It belongs to it only by its results and practical applications, but not by its inspiration; for in science, as in art, all true novelty is the work of genius; and true genius, unlike talent, is supernatural.[52]

Such reflections arise from a set of convictions very different than those shared by Rhees, Pascal, and arguably the majority of contemporary persons who contemplate these issues.

Clearly, Rhees' problem with interpreting Weil does not arise from his being "stupid" or unable to grasp her words. He understands all her words, but within his framework of assumptions he fully understands none of them. Rhees' problem with Weil is on a *metaphysical* level and is an example of an interpreter attempting to understand a text from within the confines of a particular metaphysical framework, when the author of the text is writing from within the confines of an entirely different framework.[53] It is entirely the case that Weil ultimately ties science, mathematics, and religious insights together, as we shall see, but to assume from the start, as Rhees does, that it is impossible to meaningfully do so is to assume that one's own set of first principles cannot be challenged or modified.[54]

Rhees' commitments lead him to a position similar to the metaphysic underlying modern and contemporary science, a metaphysic containing a host of problems and shortcomings that were the subject of much of the previous chapter's discussion. If, as Weil has argued, the metaphysical assumptions underlying contemporary science have led to various crises of value and meaning, then it makes a great deal of sense to consider whether other metaphysical assumptions might help us avoid these crises. This is precisely what Weil is proposing to do when she draws our attention to the insights of the Pythagoreans; Rhees' difficulties arise because he is unable to imagine a metaphysic so entirely different than the one he is steeped in.

Choosing between incompatible metaphysical worldviews is not a matter of comparing facts or devising better and better experiments; "Science has not the right to place in doubt primary hypotheses; that is not its role; it is below this level."[55] By definition, metaphysical commitments are more foundational than and prior to factual analysis and experimentation, since one's metaphysical commitments serve to define what is a fact in the first place. Despite the spectacular successes of contemporary science on the level of pragmatic applications, Weil has argued that it has become disconnected from other crucial features of human existence, including the need for value and meaning. The vacuum created by this disconnect has been filled by power and manipulation, fueled by technology. In looking back to the Pythagoreans and the Greeks, Weil is asking us to consider carefully a different way of thinking about science and reality at large, one arising from a metaphysic entirely different than the one that supports contemporary science.

Weil frequently argues that the Greeks had the knowledge and capacity to produce many more practical advances in science than they apparently did produce. This is because they consciously chose *not* to focus on practical and technological advances, understanding the perils of power and technology disconnected from values. "The disdain shown by the Greeks for the applications of science did not spring from an aristocratic turn of mind, but from this elementary truth that the applications can just as easily be bad as good."[56] Such a "disdain" is, of course, reflective of an orientation quite foreign to that of modern and contemporary science, an orientation toward something attractive but unknown. "Greek science was based on piety. Ours is based on pride."[57]

What increasingly attracted Weil to the Greek understanding of science and mathematics was, first and foremost, the possibility that science and mathematics need not be divorced from the larger scope of human activity, that these disciplines can truly be integrated with art, philosophy, religion, in short with all human intellectual and creative endeavors. In her Marseilles notebooks, she writes

At one time, I found it difficult to understand how art and science could be reconciled. Today, I find it difficult to understand how they can be distinguished. . . . The theory of beauty in the arts and the contemplation of beauty in the sciences—these two things must coincide through some hitherto unexplored path.[58]

We find, with the Greeks, proof that exploration of this path has already begun. The very existence of Greek science and mathematics, according to Weil, is due to their conviction that all phenomena, human and otherwise, point to something greater than ourselves, a possible source of meaning and value. Weil's philosophy of science and mathematics is, ultimately, an attempt to make this conviction the starting point for a new metaphysic of science.

MATHEMATICAL BRIDGES

One of the most important concepts in Simone Weil's entire philosophy is revealed by her frequent use of the Greek word *metaxu*.[59] The word literally means "between" or "intermediary"; Weil uses it in reference to things that serve as intermediaries or bridges between human beings and the divine. They are means to an end; ultimately, she will argue that much of what we consider reality (the external world, things, etc.) has meaning as a vast network of means or intermediaries toward ultimate truth and reality. For our present purposes, it is important to consider the notion of *metaxu* because in Weil's thought the notion of intermediary or bridge is most clearly exemplified by mathematics.

In Weil's estimation, the ancient Greeks had a profound understanding of the need for mediation; indeed, is was this understanding that motivated much of what we honor most of their civilization.

Greece . . . had a revelation of her own: it was the revelation of human misery, of God's transcendence, of the infinite distance between God and man. Haunted by this distance, Greece worked solely to bridge it. That was what made her whole civilization. Her mystery religion, her philosophy, her marvellous art, that science which was her own invention, and all the branches of it, these were all so many bridges between God and man.[60]

At the very core of this revelation is the need for mediation, for bridges that will serve as possible passageways from human imperfection to divine perfection. Indeed, Weil frequently suggests that the Greeks "invented the idea of mediation,"[61] seeking for evidence of *metaxu* particularly in the proportions and harmonies in the world through mathematical relationships. Balance and harmony are signposts of mediation, causing the Greeks to seek for evidence of proportion everywhere, both outwardly and inwardly. "It is for the Greeks that mathematics was really an art. It had the same purpose as their art, to reveal palpably a kinship between the human mind and the universe, so that the world is seen as 'the city of all rational beings.'"[62] Weil is also convinced that it is particularly through mathematics and science that Greeks found paradigms for mediation, balance, and proportion.

> Among the Greeks, epic poetry, drama, architecture, sculpture, the conception of the universe and the laws of nature, astronomy, mechanics, physics, politics, medicine, and the idea of virtue all have at their center the idea of balance that goes with proportion, the soul of geometry. With this idea of balance, which we have lost, the Greeks created science, our science.[63]

It is this understanding of science that Weil seeks to restore in our age: "We must give back to science its true destiny as a bridge leading toward God."[64]

Despite apparent evidence to the contrary, Weil frequently expresses optimism and confidence that a reorientation of science from its inadequate contemporary metaphysical framework to a framework that would position science in its proper relationship and role with respect to human concerns is not only possible but achievable. "Not much would be required (yet a lot in a certain sense) to bring us back from contemporary science to an equivalent of Greek science."[65] What must be recovered is an awareness of what science, mathematics, indeed all intellectual and creative human activities are *for*. We have made a mistake that is the opposite of the mistake Kant warns us of in his moral philosophy. Kant instructs us not to mistake ends in themselves for means; with respect to science, Weil maintains, we have mistaken means for ends in themselves.

> The bridges of the Greeks. We have inherited them. But we do not know what use to make of them. We have imagined that they were for building houses upon. So we have erected sky-scrapers thereon to which we are continually adding fresh storeys. We do not realize that they are bridges, things made to be crossed over, and that is the way leading to God.[66]

One could consider any one of the disciplines mentioned above—poetry, drama, architecture, sculpture, natural science, geometry, astronomy, mechanics, physics, politics, medicine, to name a few—and argue that marked developmental progress has been achieved over the last two millennia by building on the solid foundations laid in antiquity. Weil does not deny this. The problem is that the Greeks conceived of these disciplines as means, as bridges, as *metaxu* leading to an end entirely different than themselves. Subsequent generations, up to the present, have mistaken bridges for foundations upon which structures are to be built and refined as ends in themselves. "We have inherited all of them. We have built them up much higher. But we now believe that they were made to live in. We are unaware that they are only there to be passed across; we do not know, if we crossed over, whom we should find on the other side."[67] This, interestingly, provides Weil with her unique definition of *humanism*: "which consists in treating the bridges bequeathed to us by the Greeks as if they were permanent habitations."[68]

Understanding the import of Weil's distinction between bridges and permanent habitations is crucial to envisioning precisely what she is hoping for in a reoriented science. Bridges, of course, can be improved, made stronger, wider, etc. In other words, thinking of science and mathematics as bridges rather than permanent habitations does not remove the need for continued development and improvement. The progress of science, however, becomes a quite different process when seen as the improvement of a bridge rather than the construction of something to live in. If science is a bridge, then continued progress will not only improve the day-to-day existence of human beings, but will also provide better and better ways of bridging the chasm between the human and the divine. This double role is expressed clearly in the following comments concerning geometry from *The Need for Roots*:

> Geometry thus becomes a double language, which at the same time provides information concerning the forces that are in action in matter, and talks about the supernatural relations between God and his creatures. It is like those ciphered letters which appear equally coherent before as after deciphering.[69]

What are the features of mathematics that reveal its special role as *metaxu*? Weil's answer to this question will be the focus of the next chapter, but a few general comments are appropriate here. The mathematical relationships and structures underlying acoustical phenomena such as musical harmonies are one example of why mathematics is especially appropriate as a bridge, for these relationships are discovered, not invented. As Weil writes, there is a "mysterious appropriateness existing in mathematics,"[70] something unexpected and beautiful in the fact that the very fabric of the world is mathematical. It is to this "mysterious appropriateness" that Subrahmanyan Chandrasekhar refers when he speaks of "this incredible fact that a discovery motivated by a search after the beautiful in mathematics should find its exact replica in Nature."[71] For the Greeks, and for Weil, this "incredible fact" is explained by realizing that "eternal mathematics . . . is the stuff of which the order of the world is woven."[72] This appropriateness is not an accident, but directly points toward something more.

> The universe provides these images thanks to a divine favor accorded to man which allows him to make use of number in a certain way as intermediary, in Plato's terms, between the one and the unlimited, the indefinite, the indeterminate—between unity, as man is able to conceive it, and everything that opposes his attempt to conceive it.[73]

The beauty of the world is both a result of its mathematical structure and an open field for future investigation in the search for even more proportions, harmonies, and relationships. "Not only is this visible world unexceptionably beautiful, but as we proceed to study it scientifically, it reveals itself to be an inexhaustible source of beauty."[74] No wonder, Weil says, the ancient Greeks "searched everywhere—in the regular recurrence of the stars, in sound, in equilibrium, in floating bodies—for proportions in order to love God."[75]

Recognition of the mathematical structure of reality immediately points our attention beyond us, because it is abundantly clear that this structure is entirely independent of human energies, capacities, and concerns. "What is beautiful in mathematics is that which makes abundantly clear to us that they are not something which we have manufactured ourselves."[76] We continually encounter proportion, order, and harmony in the world, but we do not know why. The more we discover, the more we become aware of our own limitations. Mathematics serves as a bridge to what is other, greater than ourselves.

> Mathematics alone makes us feel the limits of our intelligence. For we can always suppose in the case of an experiment that it is inexplicable because we don't happen to have all the data. In mathematics we have all the data, brought together in the full light of demonstration, and yet we don't understand. We always come back to the contemplation of our human wretchedness. What force is in relation to our will, the impenetrable opacity of mathematics is in relation to our intelligence. This forces us to direct the gaze of our intuition still farther afield.[77]

The Greeks, according to Weil, "invented the method of rigorous demonstration" because "they perceived a divine revelation in geometry."[78] The certainty, precision, and even unexpectedness of mathematical relationships and proportions are non-human and otherworldly in a real sense. Because of this, the Greeks sought to introduce a rigor into their mathematics and science that would, as much as possible, reflect the certainty and unchanging nature of what they, as *metaxu*, point toward. "The Greeks believed that only truth was worthy to represent divine matters, not error or approximation, and the divine character of anything only made them more exacting in regard to precision, not less so, deformed as we are by our habit of propaganda."[79] Weil argued in her discussion of the "scientific village" that our contemporary tendency is to be attracted to contingency, open-mindedness and probability, rejecting the importance or even the need for certainty and rigor. The Greeks, however, sought certainty and rigor, as exhibited in geometry and mathematics first and foremost, because these are reflective of divinity. The bridge-like nature of mathematics is revealed in this reflection.

Finally, mathematics serves as a paradigmatic bridge because of the moral lessons and values that it instills in us. The moral importance of mathematics will be an important feature of our ongoing discussion; the following striking example from Weil shows just how seriously she takes the connection between mathematics and moral values:

> There is an analogy between the fidelity of the right-angled triangle to the relationship which forbids it to emerge from the circle of which its hypotenuse is the diameter, and that of a man who, for example, abstains from the acquisition of power or of money at the price of fraud. The first may be regarded as a perfect example of the second.[80]

Rush Rhees uses this passage from "The Pythagorean Doctrine" to illustrate the problem he has with Weil's insistence on blurring distinctions that he takes as obvious and given.

> To speak of *fidélité* in both these cases is an example of Weil's use of a simile or metaphor with no clue as to what the precise meaning could be. It leaves on one side the obvious *differences* between "impossibility" in mathematics, and "impossibility" in moral alternatives.[81]

In Weil's estimation, however, this is more than a mere metaphor or simile. One of the most important moral truths revealed by mathematics is that much of the moral life and virtue requires a loosening of the "avaricious tentacles of the self"; the precision and certainty of mathematics assist us in this because in mathematics we are forced to grapple with something that is not "up to us." There is "No 'I' in numbers."[82] Nowhere does Weil claim that morality and ethics can be reduced to mathematical formulae and definitions, as Spinoza might have claimed.

> It is in so far as they constitute a good, a value and end, that mathematics are "a shadow, but a divine one, an image of that which is." To want to understand a mathematical theorem isn't the same thing as to want the good; but it is closer to wanting the good than to want money is.[83]

Above all, paying attention to these features of mathematics helps us to overcome our all-too-common hubristic tendency to attempt to define the world in our own image, a tendency as old as humanity itself. "Protagoras said: *Man is the measure of all things.* Plato replies: *Nothing imperfect is the measure of anything* and *God is the measure of all things.*"[84] In no instance is the encounter with mathematical certainty wasted, even if the encounter is ultimately unsuccessful.

> It does not even matter much whether we succeed in finding the solution or understanding the proof, although it is important to try really hard to do so. Never in any case whatever is a genuine effort of the attention wasted. It always has its effect on the spiritual plane and in consequence on the lower one of the intelligence, for all spiritual light lightens the mind. If we concentrate our attention on trying to solve a problem of geometry, and if at the end of an hour we are no nearer to doing so than at the beginning, we have nevertheless been making progress each minute of that hour in another more mysterious dimension. Without our knowing or feeling it, this apparently barren effort has brought more light into the soul.[85]

For these reasons and more, Weil considers the Greek understanding of mathematics to be an essential ingredient of a scientific reorientation in our time. Mathematical and geometrical proportions are found throughout our world; the more we seek and identify them, the more we come to understand that the very existence of these relationships speaks to us of the grace that is at the heart of human existence.

> The grace which permits wretched mortals to think and imagine and effectively apply geometry and to conceive at the same time that God is a perpetual geometer, the grace which goes with the stars and with dances, play, and work is a marvellous thing; but it is not more marvellous than the very existence of man, for it is a condition of it.[86]

CHAPTER FOUR

The Divine Poetry of Mathematics

In a general way, and in the widest sense, mathematics, including under this name all rigorous and pure theoretical study of necessary relationships, constitutes at once the unique knowledge of the material universe wherein we exist and the clearest reflection of divine truths. . . . It is this same mathematics which is first, before all, a sort of mystical poem composed by God himself.

—Simone Weil, "The Pythagorean Doctrine," *IC,* 193

In the early 1930s, while she was teaching philosophy at various lycées throughout France, Simone Weil thought carefully about pedagogy in the field of mathematics, as she described in her short 1932 essay draft entitled "The Teaching of Mathematics."[1] Even in this early document, her concerns about contemporary attitudes toward mathematics are apparent, as also are her germinal ideas about how to transform these attitudes practically. Weil suggests that in the contemporary world "mathematics is nothing more for the physicist than a language enabling him to express conveniently the results of experiments"; according to one contemporary scientist, "the value of a mathematical proposition is established in the same way as that of a pass in football."[2] From a perspective now familiar from our previous considerations, Weil remarks that such an attitude toward mathematics "is the sign of a cultural decadence, a decadence symptomatic of the régime in which we live."[3]

Weil goes on to briefly describe, by way of practical example, her own "experiment" in teaching mathematics as illustrative of an entirely different way of thinking about and training others in mathematics. Her concern, as always, is to reconnect science and mathematics with real human concerns; her syllabus places us once again in the framework she believed necessary for this reconnection, a framework with far greater implications than simply a better way to teach mathematics.

> As a lecturer in philosophy I took advantage of the fact that the programme includes an examination of "method in mathematics" to devote a dozen hours to the history of mathematics, showing it to be oriented towards a solution of the fundamental contradiction between the continuous and the discontinuous (number).[4]

In the sketch that follows this passage, Weil briefly describes how these "dozen hours" devoted to the history of mathematics in her course began with Greek geometry, particularly the discoveries of Thales, Pythagoras, Eudoxus, and Menaechmus which, in her estimation, laid the groundwork for the mathematics of the Renaissance and Scientific Revolution. A close consideration of her reflections on mathematics and science over subsequent years reveals her deeper and deeper conviction that these foundational discoveries in geometry were energized by metaphysical attitudes and beliefs that, although largely unfamiliar to contemporary thinkers, are sorely needed in our age. It is to an examination of these issues that we turn in this chapter.

As we have already seen, Weil believed that the primary energy behind Greek mathematics was closer to what we would call religious and aesthetic than anything approximating our contemporary pragmatic scientific energies. As we consider the seminal discoveries of Greek geometry, we must not lose sight of that fact that the Greeks "searched everywhere . . . for proportions in order to love God."[5] Weil was fond of quoting the passage from the

Gorgias in which Socrates scolds Callicles for failing to appreciate the important moral lessons to be learned from mathematics:

> Whoever concedes to his desires cannot be in close friendship ei-
> ther with another man or with God; for he is not capable of commu-
> nion; and for whoever there is no communion, there is no friendship.
> And according to the sages, Callicles, it is communion which unites
> the heavens and the earth, the gods and men, in friendship, order,
> restraint, justice. Because of that, this universe has been named
> order (cosmos) and not disorder and licentiousness. But it seems to
> me that you do not give your attention to these things, although you
> are instructed. You have not observed how great a power geometrical
> equality is both among the gods and among men. You believe you
> should strive to acquire. This is because you do not pay attention to
> geometry.[6]

Weil is convinced that "present-day mathematics [is] a screen between man
and the universe (and therefore between man and God, conceived in the
Greek manner) instead of a contact between them."[7] If we are, then, to fully
recognize the import of the larger implications of the Greek conception of the
mathematical ordering of the world, we must avoid the mistake of Callicles
and, with Simone Weil, "pay attention to geometry."

DEMONSTRATION AND PROPORTION

The investigation of the work of ancient Greek mathematicians raises
numerous difficulties when one attempts to assign discoveries accurately and
to organize the development of theorems in the proper order, as well as when
one speculates concerning what the possible motivations behind these discoveries and theorems might be. Simone Weil was, of course, fully aware of
these difficulties. Despite her tendency to discuss such matters in a manner
containing more certainty and conviction than seem warranted, she often
admitted that her interpretation of Greek mathematics was, indeed, an
interpretation. Yet, her interpretation seemed to her to be the only one which
synthesized the various available texts in a fully coherent manner.

These comparisons may appear arbitrary, but they confer perfect coherence and intelligibility upon texts which, if I mistake not, can ac-
quire it only in this way. And there is no other criterion for piecing
together a fragmented mosaic. The sole alternative to this interpretation is to concede that the Greeks wrote incoherent and unintel-
ligible things. And that is what people have done up to now, but they
were wrong. The mistake was to judge the Greeks as if they were
like ourselves.[8]

Our failure to take the metaphysic underlying Greek mathematics seriously,
Weil contends, is a result of our viewing their thought from the perspective
of our own contemporary metaphysical framework. As discussed in the previous chapter, this is a serious mistake. To learn from the Greeks, we must
attempt to "unlearn" our own assumptions and prejudices.

One of the beginning steps in this "unlearning" process is to recognize
the rigor and precision evident in the demonstrations of Greek mathematics
and geometry. Contemporary thinkers admit that our own sciences and
mathematics are built on the foundations laid at least partially by the Greeks,
but we assume that the development of these disciplines since ancient times
has essentially been one of uninterrupted improvement both theoretically
and practically. Weil argues, however, that a careful investigation reveals
something different.

The Greeks possessed a science which is the foundation of our
own. It comprised arithmetic, geometry, algebra in a form peculiar
to them, astronomy, mechanics, physics and biology. The sum total
of knowledge accumulated was naturally very much less. But by its
scientific character, according to the full significance we attach to
that term, and judged by what we hold to be valid standards, that
science equaled and even surpassed our own. It was more exact,
precise, rigorous. . . . This science, which was as scientific as our
own, if not more so, had no trace of materialism about it. What is
more, it was not a subject of profane study. The Greeks regarded it
as a religious subject.[9]

Our tendency might be to think that religion and demonstrative rigor are di-
ametrically opposed. Weil's argues that, on the contrary, it was precisely the

religious character of mathematical inquiry that caused the Greeks to literally "invent" the notion of scientific and mathematical demonstration. "[The]
invention of demonstrative proof (the soul of our science, including the experimental method) [was] due to the Greeks' need of certainty in regard to
divine things, even in the images of them."[10]

As suggested in the previous chapter, the Greeks believed that certainty reflects divinity; hence, the certainty exhibited by mathematical proportions and relationships immediately put them in mind first of the divine.
To illustrate, Weil quotes Philolaus, whom she describes elsewhere as believing that "mathematical truth was originally theological."[11] Philolaus suggests
that it might surprise some to learn that mathematics and number can be
used for something *in addition to* theological reflection: "One can see what
powerful effect nature and the virtue of number has not only in religious and
divine things but everywhere in human acts and reasonings both in the working of various techniques and in music."[12] If mathematics is directly reflective
of divinity, then arguments and demonstrations in mathematics must be
crafted with special care; indeed, "only such a mystical conception of mathematics as this was able to supply the degree of attention necessary in the
early stages of geometry."[13] In other words, it is possible that the rigor of
mathematical demonstration that we not only take for granted but also mis-
takenly assume to be features only of modern and contemporary science and
mathematics is a direct result of the Greek refusal to separate religion and
mathematics. "It was because their mathematics was a theology that they
wanted a certainty."[14] That this sounds so odd to our contemporary ears is an
indication of just how much we have lost.

> Today we can no longer conceive this because we have lost the idea
> that absolute certainty belongs only to divine things. . . . Our intel-
> ligence has become so crude that we no longer conceive that there
> could be an authentic, rigorous certainty concerning the incompre-
> hensible mysteries.[15]

Proportions, of course, are links or bridges between things that might
appear to have little or nothing in common. If mathematics reflects divinity,
then it makes sense to imagine that a search for mathematical regularities
and proportions in the world is literally a search for intimations or traces of

divinity. Although the sequence of discoveries is always in doubt, it is possible to suppose that the unexpected discovery that mathematical relationships and proportions serve as the structure of musical harmonies was at least partially the driving force behind a concerted effort to discover such relationships in other phenomena. Furthermore, the importance of proportion and harmony as underlying features of the physical world might easily be imagined as equally important in areas of human reality that would, from our contemporary perspective, appear to be entirely outside the realm of mathematical analysis. If mathematical concepts serve as bridges in the natural world, then "one could hope to be able also to apply the notion of ratio to psychological and spiritual matters."[16]

A series of letters to her brother André, written in 1940, provides one of the most important sources of Simone Weil's mature thinking on Greek mathematics and science, including her reflections on what might have been the value of and motivations behind the Greeks' obsessive interest in proportions. She suggests to her brother that there is incontrovertible evidence "that from a fairly remote antiquity the idea of proportion had been the theme of a meditation which was one of the chief methods, and perhaps the chief method, of purifying the soul."[17] As discussed in the previous chapter, the Greeks were intensely aware of "the infinite distance between God and man";[18] as she described to her brother, "the Greeks experienced intensely the feeling that the soul is in exile."[19] At the same time, the possibility of becoming "at home" in the place of exile is very real, if only we can understand how. "This place of the soul's exile is precisely its fatherland, if only it knew how to recognize it."[20]

Proportions, mathematical relationships, and harmonies are helpful in producing this recognition. Not only do they point us toward what is other than ourselves, but they also provide us with links both direct and analogous to that other.

> Proportion enables thought to grasp all at once a complex variety in which, without the aid of proportion, it would lose itself. The human soul is exiled in time and space, which rob it of its unity; all the methods of purification are simply techniques for freeing it from the effects of time, so that it may come to feel almost at home in its place of exile. The mere fact of being able to grasp, all at once, a

multiplicity of points of view concerning one and the same object makes the soul happy.[21]

Most fundamentally, the pursuit after and discovery of proportions in all aspects of reality is a pursuit that is, in itself, crucial to becoming at home in exile.

> Thought . . . aspires to conceive the world itself as analogous to a work of art, to architecture, or dance, or music. For this purpose it is necessary to find in the world regularity within diversity; in other words, to find proportions. It is impossible to admire a work of art without thinking oneself, in a way, its creator and without, in a sense, becoming so; and in the same way, to admire the universe as if it were a work of art is to become, in a manner, its creator. And this leads to a purging of the passions and desires related to the situation of one little human body within the world; they become meaningless when thought takes the world for its object. But proportion is indispensable for this result, because without it there can be no equilibrium between thought and the diverse, complex, and changing material of the world.[22]

These, in Weil's estimation, are just some of the reasons that the Greek passion for proportion and harmony must be investigated as something far more important than a passing aberration. The energy behind Greek mathematics and geometry is nothing less than the energy necessary for human flourishing. "In the eyes of the Greeks, the very principle of the soul's salvation was measure, balance, proportion, harmony; because desire is always unmeasured and boundless. Therefore, to conceive the universe as an equilibrium and a harmony is to make it like a mirror of salvation."[23]

Conjuring Up the World

With these considerations in mind, we now turn to some of the Greek discoveries in geometry that Weil considered to be most instructive and illuminating for our present purposes. Her notebooks and published writings are filled with comments and discussions related to the work of figures such as Thales, Anaximander, Pythagoras, Eudoxus and Menaechmus, men "drunk on geometry,"[24] for whom mathematics "conjured up the world."[25] Recall that in one of her earliest texts Weil proposed that "the advent of the geometer Thales was history's greatest moment."[26] It is to this shadowy figure that we turn first.

Measuring Pyramids

In "A Sketch of a History of Greek Science," Weil writes that "Greek science had its beginning in the idea of similar triangles attributed to Thales."[27] Although it is by no means clear precisely what discoveries can be legitimately attributed to Thales, the measurement of the height of pyramids is one that ancient commentators almost universally credited him with. This measurement, in its simplest form, requires no subtle geometrical knowledge; Thales need only have made the empirical observation that at the point in the day when the length of the shadow of one object coincides with its height, then the same will be true for all other objects. As illustrated in figure 4.1 below, the height of the pyramid is equal to the shadow it casts when measured at the same time of day that the shadow cast by Thales is equal to Thales' height, or $D = E$ when $A = B$.

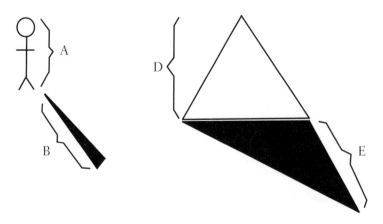

Figure 4.1

Since *A, B,* and *E* are directly measurable quantities, height *D* of the pyramid is easily established. Some commentators suggest that in his measurement of the height of the pyramid, Thales was getting close to the idea of similar triangles, a suggestion that Weil takes seriously: "It is said that the notion of similar triangles permitted Thales to measure the height of the Egyptian pyramids by their shadow and the relationship or ratio between the height and the shadow of a man at the same hour."[28] In other words, if we imagine *A* and *B* to be two sides of a triangle with imaginary side *C* drawn from the tip of Thales' shadow to the top of Thales' head, and *D* and *E* to be two sides of another triangle with imaginary side *F* drawn from the tip of the pyramid's shadow to the top of the pyramid (figure 4.2), then by the rules of similar triangles, *D* can be calculated even when *A* and *B* are not equal. The *ratio* of *A* to *B* is equal to the *ratio* of *D* to *E*; since *A, B,* and *E* are always directly measurable, *D* can easily be established.

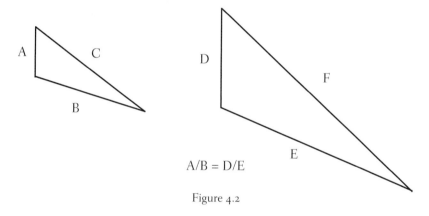

Figure 4.2

The importance of this discovery is far-reaching. First, as Weil writes, this measuring process introduces us to a geometrical, visual example of the notion of function that is so important in Greek mathematics and geometry. "Two equal ratios [*A/B* and *D/E*] which vary while remaining equal—this is the idea of function."[29] Function links constancy and variability together into a mathematically expressible relationship. Second, this discovery provides a geometrical solution to the problem of "proportion between four terms of which one is unknown."[30] As pictured above, the equation *A/B* = *D/E* can be solved for any of the four values as long as the other three values are known by constructing similar triangles. Finally, Thales' discovery is a breakthrough in the capacity of human beings to grasp their world. "Thus proportion makes things measurable, and there follows, in a sense comprehensible for man, the forbidden dimension, that would lead to heaven: height."[31] The irreducible mystery of the mathematical structure of the universe serves as a bridge between the human and divine.

> The fact that we should be able to make such things [measuring devices] and use them on the assumption that, barring accidents, they are immutable, use them while thinking, in place of them, of spheres, circles, planes, points, straight lines, and angles, and in this way use them effectively—this fact is a grace as extraordinary as the existence of the stars. It is one and the same grace, and strange to say, the object studied by science is nothing other than this grace.[32]

Similar Triangles and Mediation

Most of us remember struggling with the various proofs for the similarity of triangles that we learned in junior high or high school geometry class. In the minds of Greek geometers, Weil suggests, similar triangles provided an almost inexhaustible source of proportions reflective of God, who, as Plato said, "is an eternal geometer."[33]

> For the Greeks the construction of similar triangles, which is the foundation of geometry, was a method of seeking proportions, and doubtless the construction of the right-angled triangle, a combination of similar triangles, was a method of seeking geometric or proportional means.[34]

Weil admits that there is no way of knowing the sequence of discoveries in geometry that arose from the investigation of similar triangles, but she suggests to her brother that they may have arisen from trying to solve "the problem of finding the geometric mean between two quantities."[35]

As we saw in the previous section, geometrical discoveries attributed to Thales made it possible to find the unknown term in a proportion of four terms when the other three terms are known. There is a difference, however, between investigating proportion and mediation. While discovering how to find the missing number in a series of four proportionals is important, it is but a step. Weil speculates that "Thales . . . in studying it, intended to facilitate the study of mediation (that is to say the proportion of three terms: a/b = b/c)."[36] We have already encountered mediation in the discovery of the various mean proportionals that underlie the musical scale. Weil believed that it was in the search for how to construct such means between any quantities whatsoever that the most remarkable discoveries of Greek geometry were revealed, often in an unexpected way.

Consider once again two similar triangles (figure 4.3).

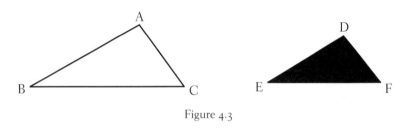

Figure 4.3

In the above similar triangles, the sides are in the proportion AB/DE = BC/EF = AC/DF. In an attempt to tease out further proportions from these relationships, Weil speculates that mathematicians might have asked, "what if the length of AC in the first triangle happens to equal the length of side EF in the second triangle?" In order to investigate the possibilities, rearrange the triangles so that they share the equal sides AC and EF, as in figure 4.4.

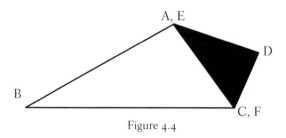

Figure 4.4

In the above similar triangles ABC and DEF, the relationships BC/EF = AC/DF = AB/DE continue to hold. Since AC = EF, BC/AC = AC/DF, or BC/EF = EF/DF. The shared line between the two triangles is the mean between *BC* and *DF*. As Weil puts it, "two similar triangles having two non-homologous [non-equal] sides represent a proportion between three quantities."[37] More generally, the investigation of the properties of similar triangles to this point has shown geometrically how to find the mean proportional between any two quantities, when the quantities are represented by sides of similar triangles.

Suppose that we now arrange similar triangles with a shared side in such a way that the extremes (*BC* and *DF*) are on the same straight line, retaining the shared side (*AC*, *EF*) as the mean proportional between *BC* and *DF*. One can visually imagine that line *DF* is rotated clockwise around point *C* until *BCD* is a straight line. Weil describes this construction to her brother as follows: "If the two extremes are constructed on the same straight line the figure becomes a right-angled triangle (because the angle between *BC* and *DF* becomes 180 degrees, whose half is a right angle)."[38] The resulting figure is represented by figure 4.5.

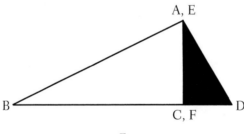

A, E

B   C, F   D

Figure 4.5

The juxtaposition of similar right-angled triangles *ABC* and *DEF*, as specified above, has produced a third right-angled triangle, *ABD*, which is similar to the two triangles of which it was constructed. As Weil writes, "The essential property of the right-angled triangle [*ABD*] is that it is formed by the juxtaposition of two triangles [*ABC* and *DEF*] similar to it and to one another."[39]

These properties of right triangles provide the information necessary to solve a number of related geometrical problems concerning mediation and proportion, problems which Weil identifies in various writings. For instance, "if the problem is set: to construct a triangle which may be divided into two similar triangles having one common side, one arrives at a construction of the right-angled triangle."[40] The solution to this problem "immediately gives the so-called theorem of Pythagoras (the sum of the squares of the other two sides is equal to the square of the hypotenuse) [$BD^2 = AB^2 + AD^2$]"[41] (figure 4.6).

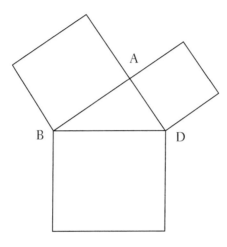

Figure 4.6

Furthermore, figure 4.5 tells us that "the height [AC] of the right-angled triangle [ABD] is the geometric mean between the segments [BC and DC] on the hypotenuse [BD]."[42] In short, we now know that the line (AC) drawn from the right angle of a right-angled triangle (ABD) perpendicular to the hypotenuse (BD) will be the mean proportional between the segments created on the hypotenuse (BC and CD): $BC/AC = AC/CD$. To find the mean proportional between any two magnitudes (BC and CD), we need simply to construct a right-angled triangle with the hypotenuse (BD) equal to the sum of the two magnitudes; the perpendicular (CB) drawn from the point on the hypotenuse at which the two magnitudes meet [C] will be the mean proportional between the two magnitudes. Since all the information needed to establish the mean proportional between any two magnitudes is contained within the right-angled triangle, the crucial problem to be solved is now: "To construct a right-angled triangle, given the hypotenuse and the position of the foot of the perpendicular."[43] The solution of this problem is one in which Weil finds great significance, a discovery that produced "the note of exultation which is audible in every reference to geometry."[44]

Circles and Sacrifices

In a number of her writings, Simone Weil emphasizes the special importance of the circle in Greek thought; in *The Need for Roots,* she writes that "the circle, in the eyes of the Greeks, was the image of God. For a circle which turns upon itself is a movement leading to no change and one completely self-contained."[45] Hence, one of the most important discoveries of the Pythagoreans was that the circle is the locus of the apices of the right-angled triangles having the same hypotenuse. According to tradition, this momentous discovery prompted Pythagoras to sacrifice a bull (some say one hundred bulls) and celebrate with a religious feast.[46]

To illustrate, consider a circle with diameter *BC* (figure 4.7). The Pythagorean discovery shows that if one specifies a point *A* anywhere on the circumference of the circle and draws lines *AB* and *AC,* resulting triangle *ABC* will always be a right-angled triangle with hypotenuse *BC.* In other words, the apex of any right-angled triangle that can be constructed with hypotenuse *BC* will fall on the circumference of the circle with diameter *BC.* The circle with diameter *BC* is the locus of the apices of all right-angled triangles having hypotenuse *BC.*

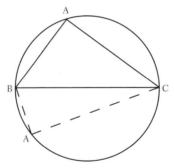

Figure 4.7

If the circle is the source of right triangles, and the various relationships contained within the right triangle are the key to finding the mean proportional between any two magnitudes, then it is not surprising that "Pythagoras offered up a sacrifice in his joy at having discovered the possibility of drawing a right-angled triangle inside a semicircle."[47] Within the context of Greek metaphysics and mathematics, this discovery reveals that mediation has a divine source.

The mean proportional was in their eyes the image of the divine mediation between God and his creatures. . . . The right-angled triangle is the source of all mean proportionals. But since it can be drawn inside a semicircle, the complete circle can be substituted for this purpose. Thus the circle, the geometrical image of God, is the source of the geometrical image of divine mediation. Such a marvelous discovery was worth a sacrifice.[48]

Weil spends a great deal of energy working out the implications of this discovery on a number of levels. In "The Pythagorean Doctrine," she asks us to reflect on what finding that the circle is the source of mediation actually entails.

When one contemplates the property which makes of the circle the locus of the apices of the right-angled triangles having the same hypotenuse, if one pictures at the same time a point describing the circle and the projection of this point upon the diameter, contemplation may extend far toward the depth and toward the height.[49]

Imagine once again, as in figure 4.8, a circle with diameter *BC,* with triangle *ABC* drawn from the endpoints of the diameter to point *A* on the circumference.

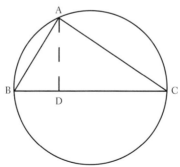

Figure 4.8

From the theorem illustrated in figure 4.7, we know that triangle *ABC* is a right-angled triangle. From the properties of right-angled triangles uncovered in the previous section, we also know that if a perpendicular is dropped from the right angle at A, intersecting hypotenuse BC at point D, resulting line AD will be the mean proportional between the two segments of the hypotenuse, BD and DC ($BD/AD = AD/DC$). The geometrical solution of the problem of finding the mean proportional between any two quantities is now complete. One need only construct a line BC out of the two magnitudes in question, BD and DC, and draw a circle with diameter BC. From the point where the segments join (D), draw a perpendicular line that will intersect with the circle's circumference at point A. Resulting line AD is the mean proportional between BD and DC.

Weil asks us to imagine point A in motion around the circumference of the circle (figure 4.9). As point A moves clockwise around the circle's circumference, Point D will move back and forth along diameter BC, with AD always remaining as the mean between the segments on the diameter.

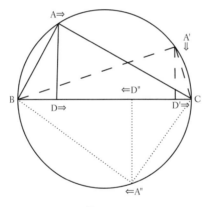

Figure 4.9

As A moves toward A' on the circumference, point D will move to the right toward point D' on the diameter; as point A moves past point C on the circumference toward point A'', point D will move back to the left toward point D''. In all cases, the various right-angled triangles retain the properties discussed above and the perpendicular from the right angle to the diameter always cuts the diameter such that the perpendicular is the mean proportional between the diameter segments.

This, Weil observes, first of all provides a striking example of the geometrical interplay of circular and linear motion, an interplay with a great deal of practical import.

> The segment on the right angle (*AD*) which joins the point of the circle (*A*) to its projection upon the diameter (*D*) is, in the figure, an intermediary between the circle and the diameter. . . . It is, like the mean proportional, the mediation between the two parts of the diameter (*BD, DC*) which are on either side of the point. . . . The affinity of the movements of the two points, one circular (*A*), the other alternating (*D*), includes the possibility of all the transformations of circular movements into alternating ones, and conversely, which are the bases of our technology.[50]

More importantly, the interplay between circle and straight line in the above figure was for the Greeks a model of the gratuitous interaction of divine and human concerns, an interaction mediated by mathematical proportions.

> In the eyes of the Pythagoreans, the element in mathematics which eludes demonstration, that is to say the coincidences, is made up of symbols of truths concerning God. It is through the proportional mean (mediation) that the passage is accomplished between circular movement (a divine act) and oscillating straight movement (a human act). All the acts of created beings are contained within impassable limits, and consequently resemble the movement of a point along a diameter, the projection on a diameter of a point which moves along a circle.[51]

There exists a mediating connection between the human and the divine that is modeled by the relationships between the circle and the inscribed right-angled triangle, "a marvellous concordance"[52] in the fact that the circle is the locus of proportional means, revealing the divine source of mediation.

Furthermore, these results reflect the tension and intertwining between *peras* and *apeiron* discussed in the previous chapter.

> To understand the range of meaning which Plato comprised under the symbolism of circular motion it should be noted that this motion is the perfect combination of whole number and continuity. . . . Circular motion is thus an image of that unison of the limited and the limitless of which Plato says in *Philebus* that it is the key of all knowledge and the gift to mortals from Prometheus.[53]

The ecstasy of Pythagoras at the discovery that the circle is the locus of proportional means is understandable when the discovery is placed within its sweeping metaphysical context.

> The circle is the image of movement which is infinite and finite, changing and invariable; surrounding an enclosed space, it evokes all those concentric circles which reach as far as the bounds of the universe; it is also, as Pythagoras with ecstasy perceived, the locus of geometric means. Circular motion obeys a law, but it has no direction; it alone is appropriate for the stars and can be attributed to them without lessening for us their power to evoke the eternal. . . . The blind necessity which constrains us, and which is revealed in geometry, appears to us as a thing to overcome; for the Greeks it was a thing to love, because it is God himself who is the perpetual geometer. From the flash of genius of Thales until the time when they were crushed by the armed force of Rome, they searched everywhere—in the regular recurrence of the stars, in sound, in equilibrium, in floating bodies—for proportions in order to love God.[54]

We shall see in the next section that the discovery of the relationship between the circle and the right-angled triangle also provided the Greeks with the necessary practical and speculative tools to address geometrically a problem that shook Greek mathematics, as well as its underlying metaphysic, to its core.

THE CRISIS OF INCOMMENSURABLES

In order to understand the import of the discovery of incommensurable magnitudes, a brief recapitulation of some of the primary Pythagorean beliefs concerning number will be helpful. As we have seen, the Pythagoreans not only established arithmetic as a branch of philosophy; they made it the basis of a unification of all aspects of the world around them. Numbers were represented by lines, each expressed as a natural number according to the ratio (*logos*) it has to another line when both have a common measure. A fraction was looked upon not as a single entity, but as a ratio or relationship between two whole numbers.

Recalling that "among the Pythagoreans one is the symbol of God,"[55] Weil writes that "The Pythagoreans considered all created things as having each a number as its symbol . . . among these numbers, some have a particular bond with unity."[56] This bond is proportion, or harmony, established between one and another number by a "mean proportional," as, for instance, 3 is the mean proportional between 1 and 9 in the relationship $1/3 = 3/9$. The numbers that have a "special bond with unity" are the numbers "which are of the second power or square," numbers such as 4, 9, 16, etc., that can be placed in proportion with unity with their square root as the mean proportional.

Aristotle ridiculed the Pythagoreans for supposedly having claimed that "Justice is a number to the second power."[57] This claim is anything but ridiculous, however, within the context of Pythagorean metaphysics. The clearest mathematical example of mediation between contraries is the relationship of x^2 to 1 by the mediating activity of x, as illustrated by the example in the previous paragraph where x equals 3. Here, Weil argues, there are intimations of something greater than mere numbers, since "the key . . . is the idea of a mean proportional and of mediation in the theological sense, the first being the image of the second."[58] As she suggests in an entry from her Marseilles notebooks, "Justice is a number raised to the second power. The just man is the one between whom and God mediation is possible."[59]

It would be a devastating discovery if it were to be found that between many whole numbers there is no mean proportional that can be expressed as a whole number. Yet this is precisely what is the case; indeed, cases such as that described in the previous paragraph are actually few and far between in the world of numbers. We have already seen this problem in our discussion of harmonic proportions in the previous chapter. In attempting to find the geometric mean between the end points of the octave under discussion, between 6 and 12, we found that the mean is specified by $\sqrt{72}$, a value that cannot be expressed by a whole number or as a quotient of integers—in contemporary terminology, it is an *irrational* number. In other words, even though "the scale . . . is symmetrically disposed around that mean," "the scale does not contain the geometric mean as a note."[60]

More generally, in attempting to establish geometric means, the Greeks discovered that the geometric mean between *any* number and its double can never be a whole number—it is always irrational. Although the mathematical proof of this is relatively straightforward, the problem of irrationality and incommensurability is best illustrated geometrically. Before proceeding, however, further definitions are in order. If one attempts to discover the geometrical mean between a number (N) and its double ($2N$), one is attempting to find X such that $N/X = X/2N$. The relationship of N/X is called *irrational* because there are no whole numbers A and B such that $N/X = A/B$; N and X do not have a ratio such as a whole number has to a whole number. By saying this, we are saying that N and X are *incommensurable*.

Even though the difficulty of establishing the geometrical mean between various numbers may have first arisen in arithmetical equations, there are strong reasons to believe that Pythagoras discovered the irrationals when investigating specific geometrical relationships. In any event, for our purposes the best example of irrationality and incommensurability arises when considering the relation between the diagonal of a square and its side, or (what amounts to the same relation) of the hypotenuse of an isosceles right-angled triangle and one of its sides, as in figure 4.10.

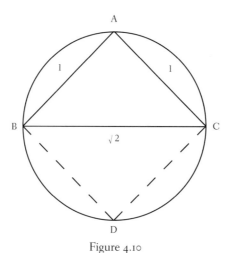

Figure 4.10

In figure 4.10, triangle ABC is an isosceles right-angled triangle ($AB = AC$) with hypotenuse BC. BC is also the diagonal of square $ABCD$. By the Pythagorean theorem, $AB^2 + AC^2 = BC^2$; hence the value of BC is $\sqrt{2}$. This, undoubtedly, was the first irrational discovered, since it is demonstrable that

BC and AB (or AC) are incommensurable. According to Weil, "Aristotle says that the incommensurability of the diagonal (BC) is demonstrated per absurdum: if it were commensurable, the even would be equal to the odd."[61] Such a proof establishes that BC and AB are incommensurable by demonstrating that the supposition that they *are* commensurable leads to an absurdity. For those readers who would like to follow the steps of the proof, they are as follows:

1. Suppose BC and AB are commensurable.
2. Then their ratio can be expressed by whole numbers; let x/y be their ratio expressed in the smallest possible whole numbers, so that $BC/AB = x/y$.
3. Then $x > y$ (since $BC > AB$), and x is necessarily >1 (since, given that y is a whole number, its smallest possible value is 1).
4. By the Pythagorean theorem, $BC^2 = AB^2 + AC^2$; since $AB = AC$, $BC^2 = 2AB^2$.
5. If $BC^2 = 2AB^2$, then $x^2 = 2y^2$ (by step 2 above).
6. Therefore, x is an even number (since any number multiplied by 2 is even).
7. Since x/y is in its lowest terms, *y is an odd number* (since if x is even and greater than y, y must be odd).
8. Suppose that $x = 2z$.
9. Therefore $4z^2 = 2y^2$ (by step 5 above), or $2z^2 = y^2$
10. Therefore y^2, and therefore *y is an even number* (since any number multiplied by 2 is even).
11. But it was shown that *y is an odd number* (step 7 above).
12. Therefore *y is both even and odd,* which is absurd.
13. Therefore the supposition that BC and AB are commensurable (step 1) has led to an absurdity; *therefore BC and AB are incommensurable.*

Subsequent discoveries established that the circumference of a circle is incommensurable with the circle's diameter and that the geometrical mean between any number and its double is irrational.

It has generally been believed that the discovery of irrationals and incommensurable magnitudes sent shock waves throughout the Pythagorean world to such an extent that all members of the Pythagorean community were forbidden from making the discovery public. Tradition has it that the first of the Pythagoreans who broke the code of silence perished in a shipwreck. Regardless of whether these stories are true, it is not difficult to imagine why the discovery of irrationals led to a crisis in the Pythagorean worldview. After the discovery of even one case of irrationality, it would be obvious that propositions that had previously been proven on the basis of a numerical theory of proportion applicable only to whole numbers were now only partially proven. The discovery of the irrationals showed that, contrary to Pythagorean beliefs, the world cannot be entirely accounted for on the basis of the relationships between whole numbers.

Since the Pythagoreans considered numbers to be of great moral and spiritual significance, the discovery of irrationals and incommensurables must have caused great distress. If at the heart of harmonic proportions, in the midst of circular motion, and in the most basic relationships of triangle sides there are irrational elements, it was no longer possible to maintain the number was the essence of all existing things or that all things were made of number. One of the reasons that the mediating activity of any number between one and its square carried great importance is because only in the case of $1/x = x/x^2$ is the geometric mean (x) guaranteed to be a whole number. The Greeks discovered that the mediating activity of x between 1 and x^2 is conspicuously missing elsewhere. In a discussion that reminds us of the great gulf existing between the human and the divine, Weil suggests that the absence of mediation between most numbers is clearly reflective of the human condition. In "The Pythagorean Doctrine," she writes that

> If one considers whole numbers, one sees they are of two sorts; those that are linked to unity by a mean proportional [that is a whole number], such as 4, 9, 16 on the one hand, and on the other hand all the others. If the first are an image of perfect justice, as the Pythagoreans say, we resemble the others, we who are in sin.[62]

The incommensurability throughout the relationships of whole numbers is reflective of the incommensurability of humanity and the divine. "Justice is that between which and God there is naturally mediation. . . . On the other hand, there is not naturally mediation between sinners and God (they are 'numbers not naturally similar'), just as there is not between unity and numbers other than square."[63]

Once the metaphysical framework that supports Greek mathematics is identified, the devastating effect of the discovery of incommensurables becomes clear. Weil suggests in various letters to her brother André that such a discovery might have led some to problematic conclusions concerning the nature of truth itself. Consider, for instance, the possible conclusions that might be drawn from the above demonstration that the geometric mean between a number and its double must be both even and odd.

> There may very well have been crisis and scandal among minds of inferior scientific and philosophic formation. . . . Who knows if the demonstration of the even being equal to the odd may not have been the model for the demonstrations proving a thesis and its contrary (the basis of sophistry) which pullulated in the fifth century and demoralized Athens?[64]

Indeed, Weil suggests, the discovery of incommensurables might have been a catalyst for the development of relativism and suspicious attitudes toward the notion of absolute truth.

> Most certainly, there *was* a drama of the incommensurables and its repercussions were immense. The popularization of that discovery brought the concept of truth into a discredit which still endures today; it brought, or at least assisted in bringing, to birth the idea that it is equally possible to prove two contradictory theses; this point of view was diffused among the masses by the sophists, along with a learning of inferior quality, directed solely towards the acquisition of power; as a result of this there arose, from the end of the 5th century, both demagogy and the imperialism that always goes with it.[65]

The situation described in this passage is remarkably similar to the situation that Weil believes to be predominant in our contemporary world, as considered in chapter 2. In the minds of imprecise and misguided thinkers, discoveries such as irrational numbers can lead to drastic problems that reach far beyond the confines of the original discipline. It may be that "the gods did right when they destroyed in a shipwreck the Pythagorean who was guilty of divulging the discovery of incommensurables."[66]

The discovery, then, of irrational numbers and incommensurable magnitudes presented Greek mathematicians with far more than a puzzle to be pondered abstractly. The entire basis of their understanding of the world around them was at stake.

> Among the Pythagoreans the words *arithmos* [number] and *logos* [reason, ratio] were synonyms. They called the irrational relationships *logoi alogoi* [ratio without reason]. To bind those numbers which are not square to unity requires a mediation which comes from outside, from a domain foreign to number which can only fulfil this function at the price of a contradiction. This mediation between unity and number is in appearance something inferior to number, something indeterminate. A *logoi alogoi* is a scandal, an absurdity, a thing contrary to nature.[67]

Greek geometers believed that their discipline was one of the primary keys to understanding reality on multiple levels; "the discovery of incommensurables seemed to destroy the very foundations of this geometry."[68] Yet, as we shall see, Weil believes that this crisis was, from the start, something to celebrate rather than to mourn, for "it [was] by force of so intense a search for a mediation of these wretched numbers that the Greeks discovered geometry."[69]

MEDIATION FROM OUTSIDE

It is in her analysis of the solution to the problem of incommensurables that we find Simone Weil at her most creative and insightful. On the one hand, as we have seen, she does not diminish the importance of the crisis. On the other hand, she argues that, far from being frozen into inaction by the appearance of irrational numbers, the "insiders" among Greek mathematicians and philosophers welcomed the challenge of incommensurability and used it as a lever to raise their metaphysical mathematics to heights it never could have reached prior to the challenge.

When reduced to its most basic elements, what exactly does the discovery of incommensurables amount to?

> The discovery of incommensurables comprises two separate discoveries: (1) that there are certain operations in integers (e.g., $\sqrt{2}$) which do not lead to any rational number and (2) that, on the other hand, these numerically indefinable results correspond to segments.[70]

Our discussion in this chapter has already established that the mean proportional between any two magnitudes can be geometrically found by making one line from the two magnitudes, drawing a circle with that line as diameter, and drawing a perpendicular from the meeting point on the diameter of the two magnitudes in question to the circumference of the circle. The perpendicular will be the mean proportional between the two original magnitudes. In the above quote, discovery (2) reminds us of this: when we consider the numbers as segments, the mean proportional can be found, whether the numbers are commensurable or not. In short, the discovery of incommensurables reveals that there are some operations that cannot be expressed numerically, even though they can be easily performed geometrically.

In letters to her brother, as well as in several essays, Weil speculates that this discovery, to those knowledgeable and insightful enough to understand, may well have been a source of joy rather than of crisis.

> He (Pythagoras) would have been overwhelmed with joy, not despair, at the discovery about the diagonal of the square. Because, to begin with, a numerical relation which cannot be numerically expressed exists nevertheless, defined by completely determined quantities. . . . If Pythagoras constructed a right-angled triangle from two similar triangles in order to form geometrical means, and if he thus obtained, what he knew he could not obtain arithmetically, the geometrical mean between a number and its double—then the note of exultation which is audible in every reference to geometry, and especially to incommensurables, is quite understandable.[71]

Greek mathematicians knew that in order to solve the problem of finding a way to express numerically what can easily be expressed geometrically, a new and different way of thinking would be required, "a purer activity of the mind, more independent of the senses, than any relation between numbers."[72] Given their conviction that mathematics and geometry are bridges between the human and divine, any breakthrough in solving the problem of incommensurables would also have great value well beyond the realm of numbers. "The Greeks, by the study of non-numerical proportions, found evidence of a much higher level, quite as exact as those in which all the terms are whole numbers. Thus they found an even more appropriate image for divine verities."[73]

We have seen that the Greeks considered the existence of geometrical relationships in the natural world as reflective of divine revelation; these intimations of divinity are brought into sharper focus when it is seen that geometry itself contains the solution to the intractable problem of incommensurables. Plato defines geometry in these terms: "Geometry . . . is the assimilation of numbers which do not by nature resemble one another, an assimilation made manifest when referred to the properties of plane figures. That this marvel is not human but divine is clear to anyone who can think."[74] Geometry could be "conceived as the theory of incommensurable magnitudes, and looked upon as a revelation."[75] Natural mediation between numbers occurs only rarely, but geometry makes it possible for there to be mediation between all numbers. "Through the destination reserved for plane figures—which is a supernatural marvel—there is mediation between unity and any number whatever. The numbers which are not naturally similar to unity are assimilated to it supernaturally."[76]

The obvious question that must have been asked at this point was whether it would be possible to refine or expand number theory in such a way as to numerically express those means that can be established geometrically. Weil expresses the problem to her brother as follows:

> The essential point of the discovery of incommensurables is outside geometry. It consists in this, that certain problems concerning numbers are sometimes susceptible of solution and sometimes insoluble; such as the problem of a geometrical mean between two given numbers. This by itself is enough to prove that number in the strict sense of the word cannot be the key to everything.[77]

Once it was discovered that "these means, whatever they may count for among numbers, have no other than a geometrical support," "it was necessary to establish that one can rigorously define the arithmetical operations and proportions of these quantities."[78]

The discovery that only by including irrational numbers under a new definition of number can the problem of incommensurables be addressed arithmetically is attributed to the great Greek mathematician Eudoxus. Essentially, Eudoxus discovered a way in which irrational lengths (such as the side and hypotenuse of an isosceles right triangle) can be compared in a way similar to the method of cross multiplying used today. His insight was that if the definition of number is expanded to cover the relative magnitude of lines, then the number system can include both rational and irrational numbers in what we would call the *real* number system. In modern notation, Eudoxus' discovery says that for any numbers a, b, c, and d, (where any or all of the numbers are possibly irrational), $a/b = c/d$ if for every possible pair of integers m and n,

1.   if $ma < nb$ then $mc < nd$,
2.   if $ma = nb$ then $mc = nd$,
3.   if $ma > nb$ then $mc > nd$.

This provides a way to compare lines or magnitudes of any length, rational or irrational. As Eric Springsted writes,

> a *logos alogos* (a ratio involving incommensurates) is not definitely expressible as we can see when we try to write the exact value of, say, the square root of two, but it is a perfectly rigorous relation between incommensurate numbers. The new theory of numbers does not in the least change the incommensuration between two numbers having no common measure, for they still have no common measure. What it does do, however, is transcend the limitations of the natural and irrational numbers by conceiving an order of numbers that incorporates both, but is not reducible to the definitions of either.[79]

Weil summarizes Eudoxus' discovery as follows:

> Eudoxus's definition of proportion, which constitutes the theory of
> generalized number, is beautiful in itself; it embraces the infinite
> variations to which four quantities are liable when multiplied two
> by two by any possible whole number, while continuing always to
> obey the law which makes these products larger or smaller than one
> another.[80]
>
> Eudoxus . . . defined with all the precision and clarity the
> human soul could desire the possibility of applying arithmetical op-
> erations to quantities not assimilable to number, such as lengths.[81]

Eudoxus' demonstration of how to compare incommensurable magnitudes was of great importance for mathematics, in that it foreshadows modern real number theory and made it possible for number theory to advance
beyond the apparent paralysis imposed by the Pythagorean discovery of irrationals. As might be expected, however, Simone Weil will argue that the true
importance of Eudoxus' discovery transcends the limitations of any single
discipline. Eudoxus has provided us with a paradigm of how apparently incommensurate features of reality can be unified on a higher level without
denying the existence of the original incommensurability.

> The notion of real number, arrived at by the mediation between
> any number and unity, was matter for just as severe demonstration,
> as clear as anything in their arithmetic, and at the same time in-
> comprehensible to the imagination. . . . By this one can conceive
> an order of certainty, starting from uncertain and easily grasped
> thoughts about the sensible world, proceeding to thoughts of God
> which are absolutely certain and absolutely inapprehensible. Mathe-
> matics is doubly a mediation between these two kinds of thought. It
> has the intermediate degree of certitude, the intermediate degree of
> inconceivability. It includes the precis of the necessity which gov-
> erns sensible things and the images of divine truths.[82]

As she writes in her Marseilles notebooks, "correlations of contraries are like
a ladder. Each of them raises us to a higher level where resides the connec-

tion which unifies the contraries."[83] And, "Since it was possible in this way to equalize the ratios in the case of two completely different pairs of magnitudes, one could hope to be able also to apply the notion of ratio to psychological and spiritual matters."[84]

The possibility of applying the thinking process behind the solution to the problem of incommensurables to areas outside mathematics is a large part of what Weil means when she suggests to her brother that "the essential point of the discovery of incommensurables is outside geometry." If it is possible to unite incommensurate numbers by referring to a higher order of numbers that would include both while preserving the incommensurability on the lower level, then the value of such an example for larger applications to a world seemingly full of incommensurate elements is beyond calculation. There is the promise of unity on the level of a transcendent order for all incommensurables, a promise worth embracing because, according to Weil, reality itself is defined by contradiction.

> Beauty is the manifest appearance of reality. Reality represents essentially contradiction. For reality is the obstacle, and the obstacle for a thinking being is contradiction. The beauty in mathematics lies in contradiction. Incommensurability, *logoi alogoi,* was the first radiance of beauty manifested in mathematics.[85]

As we saw in the chapter 2 discussion of quantum theory, Weil is not afraid of contradictions; indeed, she embraces them. Considering Eudoxus' approach to incommensurates as a paradigm for how to address contradictory features of existence, the difficulty with the contemporary "solution" of problems arising from quantum physics is clear. Eudoxus contemplated the incommensurates until he was able to identify a union on a higher level that did not ignore the basic incommensurability; contemporary scientists rushed to "solve the problem" and in doing so developed a theory that violates the most basic principles of human intuition and reason.

The insights of Greek mathematics that have been the focus of this chapter provide, in Weil's estimation, an inexhaustible source of assistance in investigating the many levels of reality, stretching from the most mundane toward the divine.

In addition to the mysterious appropriatenesses existing in mathematics, there are some still more mysterious appropriatenesses both going towards the bottom and towards the top; towards the sensible universe (in the use of mathematical ratios in the study of phenomena and in technical science); towards God (in the use of mathematical ratios as symbols of supernatural truths). It follows that the sensible universe itself, precisely in so far as it appears to our eyes as rigorously subjected to necessity, is a tissue of symbols.[86]

Eudoxus' treatment of the problem of incommensurables models the importance of learning to contemplate and face squarely matters that appear to escape the grasp of reason. Not only does such an exercise reveal the limitations of human reason, but it also brings us to the threshold of a transcendent order. "This notion forces the mind to deal in exact terms with those relationships which it is incapable of representing to itself. Here is an admirable introduction to the mysteries of faith."[87] There is no guarantee that all incommensurables can be unified in this manner; it may be that some incommensurables cannot be mediated. As we climb the "ladder of contraries," however, we may be brought to the place where, although we are no longer capable of the unification in our own power, we will be open to the possibility of supernatural unification.

Correlations of contraries are like a ladder. Each of them raises us to a higher level where resides the connection which unifies the contraries; until we reach a spot where we have to think of the contraries together, but where we are denied access to the level at which they are linked together. This forms the last rung of the ladder. Once arrived there, we can climb no further; we have only to look up, wait and love. And God descends.[88]

CHAPTER FIVE

"All Geometry Proceeds from the Cross"

In the minds of the best of the Greeks there dwelt the idea of mediation between God and man, of mediation in the descending movement by which God seeks man. . . . The idea of mediation achieved full realization, the perfect bridge was seen, divine Wisdom became visible, as Plato had hoped, to mortal eyes. In this way the Greek vocation was perfected by becoming the Christian vocation.

—Simone Weil, "The Romanesque Renaissance," *SE*, 46

At the beginning of this study, we noted Czeslaw Milosz's observation that "unlike those who have to reject their past when they become Christians,"[1] Simone Weil's thought is remarkable for its developmental continuity, deepening in insight and maturity after she turned her attention markedly in the direction of "religious" issues in the final years of her life. Her profound mystical encounters with "something stronger than I,"[2] something that "compelled me for the first time in my life to go down on my knees,"[3] in 1937 and 1938 were as unexpected as they were life changing, as Leslie Fiedler describes in his introduction to *Waiting for God.*

> Surely, no "friend of God" in all history had moved more unwillingly toward the mystic encounter. . . . The particular note of conviction in Simone Weil's testimony arises from the feeling that her role as a mystic was so *unintended,* one for which she had not in any sense prepared. An undertone of incredulity persists beneath her astonishing honesty: quite suddenly God had taken her, radical, agnostic, contemptuous of religious life and practice as she had observed it.[4]

It is partially because of the unexpected nature of these mystical encounters that Weil serves as a model of how all of a person's intellectual and emotional energies can be engaged in the attempt to graft new experiences and knowledge onto his or her existing stock of beliefs and commitments.

A careful reading of Weil's own account of the experiences in Portugal, at Solesmes, and at Assisi that transformed her indicates that in retrospect she was able to see that her whole life, both intellectually and physically, had actually prepared her not only for the encounters but also for the profound and unique way in which she came to understand the heart of Christianity. Even before these experiences, she writes, "I knew quite well that my conception of life was Christian."[5] Her debilitating headaches at Solesmes "enabled me by analogy to get a better understanding of the possibility of loving divine love in the midst of affliction . . . in the course of these services the thought of the Passion of Christ entered into my being once and for all."[6] A "chance" encounter with "a young English Catholic" man who introduced her to the poetry of George Herbert led to her using Herbert's poetry as a "prayer" that served to somewhat ease the excruciating pain of her headaches. It was during her recitation of Herbert's "Love" that Weil says "Christ himself came down and took possession of me."[7]

We have already seen that "Pascal, when he was on the point of discovering the algebraic form of the integral calculus, abandoned algebra and geometry because he desired contact with God."[8] In *The Need for Roots,* Weil argues that Pascal was motivated by fear in his decision more than by any admirable desire to move toward God.

> Pascal had already been guilty of lack of probity in his search for God. Having had his mind formed by the practice of science, he didn't dare hope that by allowing it full play it would find certitude

in the Christian dogma. He didn't dare run the risk either of having to do without Christianity. So he undertook an intellectual research having decided beforehand where it was to lead him. . . . He allowed his mind to be dominated by a conscious and deliberately entertained suggestion. After which, he sought for proofs.[9]

Weil's intellectual rigor and energy were of such power that she never considered the possibility, even after Christ "took possession" of her, that she should set her considerable intellectual abilities aside in order to pursue God, as if intellectual rigor and the divine were somehow incompatible. Instead, she sought to situate her intellect and lifelong passionate search for truth in the context of the new framework within which she found herself.

God in his mercy had prevented me from reading the mystics, so that it should be evident to me that I had not invented this absolutely unexpected contact. Yet I still half refused, not my love but my intelligence. For it seemed to me certain, and I still think so today, that one can never wrestle enough with God if one does so out of pure regard for the truth. Christ likes us to prefer truth to him because, before being Christ, he is truth. If one turns aside from him to go toward the truth, one will not go far before falling into his arms.[10]

In this chapter and the next, we will consider the development of Weil's thought in the realm of mathematics and science illuminated by her encounters with the transcendent.[11] We shall find that, just as Weil was prepared in unexpected ways for contact with the divine, so she read Greek mathematics as a preparation for the advent of the Christ. "The appearance of geometry in Greece is the most dazzling of all the prophecies which foretold the Christ."[12]

ARCHIMEDES AND THE LITURGY

Simone Weil would have rejected the claim that her later thought turned in an increasingly "religious" direction, if that claim is meant to indicate that such a turn also marked a turning away from themes and issues that had consumed her throughout her life to that point. We have seen that in her estimation disciplines as diverse as mathematics, art, politics, and theology are but distinct threads in the same cloth of human investigation and experience. Still, as we consider the manner in which she links these various disciplines under the additional influence of both her mystical experiences and Christianity, we will encounter cryptic claims that, on the surface, challenge the abilities of even the most sympathetic interpreter. What are we to make of claims such as "all geometry proceeds from the cross,"[13] "geometry, and consequently the whole of modern science, were born of faith in the Incarnation,"[14] and "the invention of geometry in Greece in the sixth century B.C. appears to be in the strictest sense of the word a prophecy"?[15]

To begin our investigation of the truly sacramental character of Weil's understanding of mathematics and science we will briefly consider some striking comments she makes concerning the famous Greek scientist Archimedes, the father of mechanics and physics. These comments illustrate the sorts of ways in which Weil came to connect mathematics and science with faith. Archimedes exemplified the passion and excitement generated by continuing discoveries in geometry and science among the ancient Greeks, as the following passage from Plutarch reveals. Interestingly, the description would also be believable if it were of Simone Weil.

> The charm of his familiar and domestic Siren made him forget his food and neglect his person, to that degree that when he was occasionally carried by absolute violence to bathe or have his body anointed, he used to trace geometrical figures in the ashes of the fire, and diagrams in the oil on his body, being in a state of entire preoccupation, and, in the truest sense, divine possession with his love and delight in science.[16]

Recall that it was Archimedes who reportedly said "give me a point of leverage and I will move the world."[17] In "A Sketch of a History of Greek Science," Weil meditates on this claim and its implications.

> Archimedes said: "Give me a fulcrum and I will lift up the world." To carry out this boast two conditions were needed. First, that the fulcrum itself should not belong to this world. Then, that this ful-

crum should be at a finite distance from the center of the world and
at an infinite distance from the hand which acts. The operation of
lifting the world by means of a lever is impossible except to God.[18]

Archimedes' bold claim is possible in the abstract, given the laws of physics
and mathematics, but impossible in reality because only something outside
the world could manipulate the lever to move the world. There is a distance
between the manipulator and the manipulated that requires a connecting
point, represented in Archimedes' claim by the fulcrum. What might this
fulcrum be?

> The Incarnation supplies the fulcrum. Every sacrament is such a
> fulcrum. And that every being who perfectly obeys God is such a
> fulcrum. For he is in the world but not of the world. . . . One may say
> that God acts here below in this manner only, that is, only by means
> of the infinitely small who, although being opposed to the infinitely
> great, are effective by means of the law of leverage.[19]

The Christian doctrine of the Incarnation, according to which God became
human in the form of Christ, both human and divine, illustrates for Weil
precisely what is needed for there to be effective contact between the divine
and human—something that participates in both. And, remarkably, Weil
does not wish us to think of this connection as a merely interesting parallel
between physics and theology. In her mind, Christ truly *is* the fulcrum that
makes it possible for the divine to touch the world.

> This is why the liturgy can, in all strictness, say that the Cross was
> a scale whereon the body of Christ acted as counterpoise to the
> weight of the world. For Christ belonged to heaven, and the distance
> of heaven to the point of intersection of the branches of the Cross is
> the distance from this point to the earth, as the weight of the world
> to that of the body of Christ.[20]

Although it is too early in our considerations of these matters to analyze these remarkable passages thoroughly, the richness of insight that Weil's
contact with Christianity brought to her consideration of mathematics and
science is manifestly clear. In the above passages, the crucial theme of

mediation takes on an even deeper significance. The Greeks believed that humanity and the divine are separated by a great gulf in need of mediation that only the divine itself can provide; mathematics provides a model of this "bridge," particularly in the discoveries concerning incommensurables. In Christianity, the divine mediator becomes incarnate, is a person both human and divine. As Weil describes in the quote that opened this chapter, in Christ "the idea of mediation achieved full realization, the perfect bridge [*metaxu*] was seen."[21] Furthermore, the above passages hint at the possibility of the Incarnation continuing now in human persons. It is possible for those who "perfectly obey God" to serve as the fulcrum through which God can act in this world.

These and similar themes will be developed throughout the course of this present chapter. As we proceed, it is necessary to keep in mind that for Weil, as for the ancient Greeks, the symbols and metaphors that we find throughout mathematics and geometry are far more than "interesting" parallels to seemingly unrelated matters. In her Marseilles notebooks, she writes that there are only three possible reasons to be interested in science and mathematics. "The practical interest of science can only lie in three directions: (1) technical applications, (2) game of chess, (3) road leading toward God."[22] Directions (1) and (2) are almost exclusively the ones taken in the contemporary world; Weil, obviously, wants us seriously to consider direction (3). As a *metaxu* toward God, the symbols of mathematics are literally the language of the divine, if we but learn to read them properly. As Weil writes, this "reading" of the world, the belief that "the world is God's language to us,"[23] has a tradition far older than our contemporary use of mathematics and science primarily for technical applications and increase of power. "The foundation of mythology is that the universe is a metaphor of the divine truths. . . . The story of Christ is a symbol, a metaphor. But it used to be believed that metaphors produce themselves as events in the world. God is the supreme poet."[24] In *The Need for Roots,* Weil summarizes the above considerations as follows:

> It is not only mathematics but the whole of science which, without our thinking of noticing it, is a symbolical mirror of supernatural truths. . . . If in the natural and mathematical sciences symbolical interpretation were to occupy once again the place it occupied

formerly—the unity of the established order in this universe would appear in all its sovereign clarity.[25]

Concerning the matters before us in this chapter, Simone Weil warns us that we must walk a fine line between reducing mystery to what can be grasped intellectually and refusing to allow the intellect to engage mystery at all. To walk this fine line properly, we must be fully aware of what, in her New York notebooks, Weil calls "the most important truth . . . the fundamental doctrine of Pythagoreanism, of Platonism, and of early Christianity." This truth is, simply, that "there are two kinds of reason."[26]

> There is a supernatural reason. It is the knowledge, *gnosis*, of which Christ was the key, the knowledge of the Truth whose spirit is sent by the Father. What is contradictory for natural reason is not so for supernatural reason, but the latter can only use the language of the former. Nevertheless, the logic of supernatural reason is more rigorous than that of natural reason. Mathematics offers us an image of this hierarchy.[27]

As such an image, mathematics provides insight into "the source of the dogmas of the Trinity, of the two-fold nature of Christ in one person, of the duality and unity of good and evil, and of transubstantiation, which have been preserved by an almost miraculous protection, one might think, of the Holy Spirit."[28] Nowhere does Weil claim that the mysteries of the Christian faith can be read directly from a deep immersion in the intricacies of mathematics. As we shall see, the metaphors are much more complex than that. Still, mathematics as a *metaxu* can illuminate the mysteries that by their very nature are beyond the reach of human reason.

> The realities of the dogmas of the Trinity, of the Incarnation and of the Passion. This certainly does not mean that the dogmas could have been found by human intelligence without revelation. But once the dogmas have appeared, they impose themselves upon the intelligence, if it is illumined by love, with such certainty that it cannot refuse to adhere, even though those dogmas are out of its domain and the intelligence is not qualified to affirm or deny them.[29]

For these and many other reasons, mathematics serves as "an admirable introduction to the mysteries of faith."[30] In the continuing search for bridges between the human and divine, we look also for the proper language in which to express these mediating links. "Natural reason applied to the mysteries of the faith produces heresy. The mysteries of the faith, when severed from all reason, are no longer mysteries but absurdities. But supernatural reason only exists in souls which burn with the supernatural love of God."[31]

MARVELLOUS MEANINGS

Toward the end of "The Pythagorean Doctrine," Weil writes that relationships such as those we find expressed in mathematical proportions are at the core of our human interaction with the world on all levels. It is in such relationships that we find intimations of the divine.

> We have in us and about us only relationships. In the semi-darkness in which we are plunged, all is relation for us, as in the light of reality all is in itself divine mediation. Relationship is divine mediation glimpsed through our darkness. This identity is what St. John expressed in giving Christ the name of the relation, *logos*, and what the Pythagoreans expressed in saying: "All is number." When we know that, we know that we live in divine mediation, not as a fish in the sea but as a drop of water in the sea. In us, outside us, here below, in the Kingdom of God, nowhere is there any other thing. And mediation is exactly the same thing as love.[32]

In this essay, Weil asks us to consider carefully the full implications of a favorite Pythagorean saying concerning friendship which, if interpreted according to the mathematical insights that we have considered in the preceding chapters, provides an important link between Greek thought and Christianity.

> The formula: "Friendship is an equality made of harmony," is full of marvellous meanings with regard to God, with regard to the communion of God and man, and with regard to men, provided that the

Pythagorean sense of the word harmony is taken into account. Harmony is proportion. It is also the unity of contraries.[33]

The various "harmonies" that Weil identifies in the course of "The Pythagorean Doctrine" reveal the activity of divine mediation on many levels of reality. "God is mediation, and all mediation is God. God is mediation between God and God, between God and man, between man and man, between God and things, between things and things, and even between each soul and itself."[34]

In this chapter and the next, we will attempt to untangle some of the various webs of mediation that Weil wishes us to consider. Her remarkable study of the possibility that "it was because of their faith—faith inspired by the love of Christ—that the Greeks had that hunger for certainty which made them invent the geometrical proof"[35] will shed light to illuminate "an infinitely precious part of the Christian doctrine [that] has disappeared."[36] At the heart of this illumination is "the discovery that intoxicated the Greeks: that the reality of the sensible universe is constituted by a necessity whose laws are the symbolic expression of the mysteries of faith."[37]

God and God

There may be no doctrine of the Christian faith more difficult to comprehend than the Trinity, yet Weil claims that "The Trinity and the Cross are the two poles of Christianity, the two essential truths."[38] As the mediation between "God and God" is the first relationship discussed in "The Pythagorean Doctrine" as "full of marvellous meanings" in the context of harmony and proportion, it is to this difficult core relationship that we turn first. We shall see that it, as all true harmony and proportion, is a unique expression of divine love. "Before all things, God is love. Before all things, God loves himself. This love, this friendship of God, is the Trinity."[39]

Most basically, the doctrine of the Trinity states that God is a unity, yet is also three distinct Persons: Father, Son, and Holy Spirit. The normal human reaction to such a paradoxical claim is to find a way, a "technique," for figuring out what the claim could possibly mean; by doing so, one eliminates the discomfort generally felt when one encounters paradox and mystery. In her Marseilles notebooks, Weil describes such an attempt with reference to the Trinity.

> It is easy to conceive of three gods. . . . It is easy to conceive of a
> single God. It is impossible to conceive of the two at the same time,
> at one go. But one can conceive of the two alternately with sufficient
> rapidity to give oneself the illusion of simultaneousness. The same
> is true in the case of Christ as God and Christ as Man. The use of
> the mystery is then nil.[40]

The Trinity presents us with two ideas that are incommensurable. Something cannot be both one thing and three things at the same time. Yet we
can easily create the illusion of understanding by a technique such as Weil
describes in the above passage. Such a technique is similar to that used by
contemporary scientists when faced with a paradox such as the "discrete
packets of energy," as we saw in chapter 2.

Yet, as Weil says, such techniques not only fail to grasp the presented
mystery, they disguise or dismiss the thing of most value—the mystery
itself.

> It is as though one were to measure the height of a star above the
> horizon from two different spots on the earth's surface, but omitted
> to think on the two measurements at once by the process of triangu-
> lation. One would then learn nothing about the distance of the
> star.[41]

Weil's example is illustrated in figure 5.1. In order to measure the distance
from the earth to star A, choose two points on earth B and C and measure
the angle between the horizon and A from both locations (angles CBA and
BCA). Since the distance between points B and C is easily measurable, and
the angles CBA and BCA are known, the distance BA from the earth to the
star can be calculated according to the properties of triangles. Trying to think
simultaneously of the unity and plurality of God as expressed in the doctrine
of the Trinity would be similar to trying to measure the distance to the star
without considering the relationship that connects the two measurements
via the properties of triangles.

When one does pay attention to proportion and relationship, one can
learn something about the star even though it is far beyond our direct reach.

Figure 5.1

This teaches us a great deal about how to engage contradictions and para-doxes in general.

> Just as one takes a sight on the star with the aid of the two directions combined in the triangle, so in the same way one takes a sight on God with the aid of the two truths conceived simultaneously. It is always so when using relation. Two truths conceived simultaneously through the link supplied by relation enable us to seize hold, as with two sticks, of a point that is situated outside our direct range. But in the case of the mystery [the Trinity], the fact that it is impossible to conceive the two ideas together by means of a relation, because they are contradictory, results in the point aimed at, in other words God, being transported even beyond the infinite.[42]

There is much of instructive value here. The method of triangulation does indeed reveal the distance between the earth and the star; there is no method of "triangulation" that will reveal in a similarly clear sense the "dis-tance to God," or, in the case in question, that will reveal how God can be both a unity and plurality at the same time. Yet mathematics shows us the possibility of learning more through the concepts of proportion and relation concerning what is beyond our reach than if we did not consider what mathematics teaches us about harmony and mediation. Weil's meditations on the Trinity never stray far from the basic truth that "all is mediation," even in the realm of the transcendent relationships between God and God.

When reflecting on the Trinity, Weil most often begins with two Pythagorean sayings.

> "Harmony is the unity of a mixture of many, and the single thought of separated thinkers." We may place alongside it another Pythagorean formula: "Friendship is an equality composed of harmony." These two formulae together would make a perfect starting point for a theologian wishing to speak about love in the Trinity.[43]

These Pythagorean formulae draw our attention to the contraries of unity and plurality, contraries that are perfectly harmonized in the Trinity. The unifying energy is love:

> Between the terms united by this relation of divine love there is more than nearness; there is infinite nearness or identity . . . God is so essentially love that the unity, which in a sense is his actual definition, is a pure effect of love . . . the infinite virtue of unification.[44]

Yet this unity is composed of "separated thinkers," a plurality whose individuality is not subsumed under unification. Both unity and plurality—how can this be?

As is often the case, Weil provides us with geometrical pictures that, if considered carefully, will provide a certain amount of insight into this mystery. Just as the Greeks considered the circle to be the geometrical figure most reflective of perfection and divinity, so the circle can provide a picture of the continuing activity of the Trinity. Circular movement is "a movement which changes nothing, which curls upon itself; the perfect image of the eternal and blessed act which is the life of the Trinity."[45] As Weil writes in *The Need for Roots*, "the symbol of the circular movement expressed for them [the Greeks] the same truth that is expressed in Christian dogma by the conception of the eternal act on which is based the relationship between the Persons of the Trinity."[46] Weil even suggests on occasion that the continuing circular motion of the stars around the pole is a physical event that should be read sacramentally, pointing toward the constant activity of the Trinity. "We should read in the pole the unity of God, in the rotation of the fixed stars the eternal act of the Trinity."[47] Once we recall additionally that the circle is the geometrical source of mean proportionals via right triangles, the spiritual significance of this geometrical figure is illuminated even further, a significance that will be increasingly revealed as we proceed.

Yet there is more to the Trinity than the unity expressed by circular motion. As the passage from *The Need for Roots* above expresses, the Trinity is a "relationship between Persons." If the Pythagorean statements are to be applied to the mystery of the Trinity, then the harmony that unifies the separated thinkers can best be characterized as friendship.

> If friendship is interpreted as an equality made of harmony, using for the definition of harmony the common thought of separate thinkers, then the Trinity itself is the friendship above all in excellence. . . . "Friendship is an equality made of harmony" further includes the two relationships indicated by St. Augustine in the Trinity, equality and connection. The Trinity is the supreme harmony and the supreme friendship.[48]

Weil will use the same definition of friendship to define the most important relationships between human beings as well as the relationship between God and humanity; the friendship between the Persons of the Trinity provides a divine model that, although unreachable by human beings, is the paradigm of love-based relationships.

> Lovers or friends desire two things. The one is to love each other so much that they enter into each other and only make one being. The other is to love each other so much that, having half the globe between them, their union will not be diminished in the slightest degree. All that man vainly desires here below is perfectly realized in God. We have all those impossible desires within us as a mark of our destination, and they are good for us provided we no longer hope to fulfil them.[49]

Even within the Godhead, which is goodness and perfection in its fullness and entirety, there is the union of opposites, unity and plurality. As we consider why, despite the divine perfection, God chose to create, we move to the next pair of contraries that the notions of harmony, proportion, and mediation may assist us in understanding, that of God and things, of Creator and created.

God and Things

The Pythagoreans understood the activity of creation as consisting of the imposition of limit and order on that which is limitless and chaotic. What does the ordering in our universe is the imposition of mathematical structures and relationships on unlimited matter, as illustrated by the mathematical relationships that produce the musical scale. These Pythagorean notions can be imported directly into the framework of Christianity, since in the Christian tradition "the principle of all limitation is God. Creation is matter brought into order by God, and this ordering action of God consists in imposing limits. Indeed this is also the conception of Genesis."[50] As Simone Weil considers the dynamic of creator and creature in the context of proportion, mediation, and harmony, however, she reveals insights that go far beyond "In the beginning God created the heavens and the earth." These insights may be her most unique and profound contribution to the dialogue between reason and faith.

The act of creation is often understood as an attempt on the part of the creator to expand his or her sphere of influence or power. Weil's understanding of creation focuses on an entirely different aspect of the creative act, that in a profound way a true act of creativity reduces or diminishes the power of the creator, in that the creator must place part of his or her own being into the created object in order for it truly to exist in its own right. We touch on this aspect of creativity when we say things such as "I really put a lot of myself into that paper," or "that project took a lot out of me." When we begin trying to understand God as Creator, it becomes especially important to pay close attention to the renunciation or diminishment aspect of the creative act. For, if God is the perfection of goodness and completion, expressed in the Trinity as the supreme harmony and the supreme friendship, then for God to create something other than God requires that God impart something of the divine essence into the creation in order for it to exist independently. This, in turn, requires that God withdraw from the creation, voluntarily renouncing the full exercise of divine power and influence. Hence, Weil writes in her New York notebooks that "because he is the creator, God is not all-powerful. Creation is abdication. But he is all-powerful in this sense, that his abdication is voluntary. He knows its effects, and wills them."[51]

Perhaps the most important question to be asked first is: Why did God create at all? In Weil's estimation, the same love that binds the Persons of the Trinity in perfect friendship is the energizing force behind the divine creative act.

> Why did God create? It seems so obvious that God is greater than God and the creation together. At least, it seems obvious so long as one thinks of God as Being. But that is not how one ought to think of him. So soon as one thinks of God as Love one senses that marvel of love by which the Father and the Son are united both in the eternal unity of the one God and also across the separating distance of space and time.[52]

Creation motivated by love is a creation that must, of necessity, include the diminishment of the creator in order that the created may exist in independence and freedom. Divine love is itself a mystery, of course, but it seems that both the unity of the Trinity and the separation required between Creator and creation equally flow from divine love.

> The love between God and God, which in itself *is* God, is this bond of double power; the bond which unites two beings so closely that they are no longer distinguishable and really form a single unity, and the bond which stretches across distance and triumphs over infinite separation . . . these are two forms expressing the divine value of the same love, the Love which is God himself.[53]

God chose to create a world of creatures who are not God, whose existence is separate from and other than God's. This requires the voluntary distancing of God from the creation, a distance that represents what Weil calls "the supreme contradiction." It is through the divine mediation of this contradiction that the extent of God's love for creation is revealed in its greatest depth. "Contradiction is our path leading toward God because we are creatures, and because creation is itself a contradiction."[54]

Between the divine Creator and the creation exists, according to Weil, the greatest distance imaginable. Creation itself requires the divine imposition of limit and order on that which is unlimited and chaotic, and yet the very nature of the creative act requires the distancing of the Creator from the creation. Inert matter, which is the "stuff" that must have limit and order imposed upon it, could not be more different than or distant from the divine Creator. It is precisely of such incommensurable entities that Philolaus wrote "Things which are neither of the same species nor of the same nature, nor of the same station, have need to be locked together under key by a harmony capable of maintaining them in a universal order."[55] Truly in the case of the Creator and creation, whatever harmony there can be will be the "union of opposites."

What is this key that binds Creator and creation together? It is at this point that Weil's insights into Greek geometry and mathematics intersect with Christianity in a unique and profound way.

> Harmony is the union of opposites. The primary pair of opposites, and the one between which lies the most unfathomable gulf, is that separating the Creator from the creature. The Word is the geometric mean, the harmonious accord between this pair. That is why we have "per quem omnia facta sunt." It is the same as what Philolaus says. The union between that which determines and that which is indeterminate. Nature, matter, the creature as such—these are what is indeterminate. The determining principle is the Creator. The link, harmony, the geometric mean, is the order of the world; the Word as ordering principle. It is this which holds the Creator and the creature together under lock and key and prevents them from drawing apart.[56]

In the concept of the Logos, the Word, Weil finds the connection between Creator and creation; at the same time, this concept is precisely the same as that of the proportional mean between incommensurable quantities in geometry. "If we say that harmony, which is the union of opposites, is the same thing as proportion, which is the proportional mean, then we have exactly the idea of the Mediation of the Word."[57]

To drive her point home, Weil draws our attention to the opening sentence of the Gospel of John.

> Geometry is the science of the search for proportional means by way of the incommensurable proportion . . . it proceeds from a supernatural revelation. . . . The invention of geometry in Greece in the sixth century B.C. appears to be in the strictest sense a prophecy. The least inexact translation of the beginning of the Gospel of St. John might be: *"In the beginning was the Mediation* (Logos), *and the Mediation was with God, and the Mediation was God. The same was in the beginning with God. All things were made by it; and without it was not anything made that was made."*[58]

The reader's attention in the first chapter of this Gospel is usually drawn directly a few lines down to the passage Weil translates as "And the Mediation was made flesh and dwelt among us,"[59] the Incarnation. Weil, however, is most interested in the earlier lines of the Gospel quoted above. The mediating activity of the Logos is "in the beginning"; indeed, this mediating activity is what binds Creator and creation "together under lock and key" from the start. As Allen and Springsted write,

> Here Weil is only talking about a unique relation between the creator and the creature (that is, the creation), and not about the more specific relations between God and creatures. In short, she is concerned with a mediating relation between the contraries of God and *all*-that-is-not-God.[60]

The Incarnation, the Logos becoming flesh, will be considered in the following section. At this point, however, we can already see that the Incarnation is a special instance of the divine mediating activity that has occurred since the beginning of creation in order to bind the Creator and creation together.

In "The Pythagorean Doctrine," Weil uses the discovery that the right-angled triangle is the source of mean proportionals to establish important insights connecting mathematics and the divine creative act. Recall one of the figures from the preceding chapter:

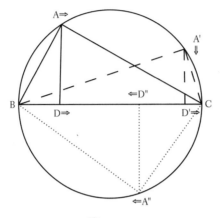

Figure 5.2

Weil asks us to imagine point A in motion around the circumference of the circle (figure 5.2). As point A moves clockwise around the circle's circumference, Point D will move back and forth along diameter BC, with AD always remaining as the mean between the segments on the diameter. Her comments on this figure, quoted partially in the previous chapter, are quoted more fully here.

> When one contemplates the property which makes of the circle the locus of the apices of the right-angled triangles having the same hypotenuse (BC), if one pictures at the same time a point (A) describing the circle and the projection of this point upon the diameter (D), contemplation may extend far toward the depth and toward the height. . . . The circular movement, if one conceives not of a point (A) but a whole circle turning upon itself, is the perfect image of the eternal act which constitutes the life of the Trinity. . . . The alternating movement of the point (D) which comes and goes upon the diameter (BC) enclosed by the circle, is the image of all becoming here below; an image made of successive and contrary ruptures of equilibrium, the equivalent variant of a motionless equilibrium in

action. This becoming is indeed the projection of divine life upon earth. As the circle encloses the moving point (*D*) upon the diameter, God assigns a term to all the becomings of this world. . . . The segment on the right angle (*AD*) which joins the point of the circle to its projection upon the diameter is, in the figure, an intermediary between the circle and the diameter. At the same time, from the point of view of quantities, it is, like the mean proportional, the mediation between the two parts of the diameter (*BD* and *DC*) which are on either side of the point. This is the image of the Word.[61]

Summarizing to this point, Weil draws our attention to the *Philebus,* where Plato writes that "the reality called eternal proceeds from the one and the many and has limit and limitlessness rooted in itself."[62] The first set of contraries, one and many, is unified in the loving friendship of the Trinity. The second pair of contraries, limit and limitlessness, is unified by the process of creation, in which "number is the intermediary between the one and the limitless."[63] The world is ordered by number and mathematical relationships, which are signs of the divine itself, as the above passage shows. The Logos is both the Word and the mean proportional, binding contraries and incommensurables together. There is more, however. In the Christian tradition, the Logos is not merely an abstract principle of organization. It is a Person, the second Person of the Trinity. "The first couple of contraries is God and the creature. The Son is the unity of these contraries, the geometrical mean which establishes a proportion between them: He is the mediator."[64] Understanding the Personhood of the Mediator causes us to also understand in a deeper way the continuing divine sacrifice involved in the divine act of creation.

Weil assumes that "there cannot be any relationship in God whose terms are not Persons, just as the bond which links them must be a Person."[65] All of the members of the Trinity are Persons, hence the unity of the Trinity is a direct expression of the above assumption. When considering the relationship between Creator and creation, however, the "Personhood" requirement creates a significant obstacle to mediating the contraries.

Since there is in God as creator a second pair of contraries, there is in Him also a harmony and a friendship which are not defined alone

by the dogma of the Trinity. There must also be in God unity
between the creative and ordering principle of limitation and the
inert matter which is indetermination. For that, not only the principle of limitation but also the inert matter and the union between
the two, must be divine Persons.[66]

The problem, of course, is that "inert matter does not think, it cannot be a
person."[67] How then, is the above requirement concerning Personhood to be
satisfied?

Weil's answer draws us to the heart of the divine love of the Creator,
not just for humanity, but for all of creation itself. For, the Personhood requirement can be satisfied only by the ultimate sacrifice.

The insoluble difficulties are resolved by passing the limit. There is
an intersection between a person and inert matter; this intersection
is a human being at the moment of death, when the circumstances
preceding the death have been brutal to the point of making a thing
of that person. This is a slave dying, a miserable bit of flesh nailed
upon a cross. If this slave is God, if he is the Second Person of the
Trinity, if he is united to the First Person by the divine bond which
is the Third Person, one has the perfection of harmony as the Pythagoreans conceived it, harmony in which is found the maximum
distance and the maximum unity between the contraries.[68]

Only the continuing suffering, agony, and death of God in the divine role
as Logos, as Mediator, can achieve the bind Creator and creation together
"under lock and key." Weil's point is that the Incarnation and Passion of
Christ are already in place at the very moment of creation, given the necessity that God's divine power be diminished in order for creation to exist. As
she writes in her New York notebooks, "Already before the Passion, already
by the Creation, God empties himself of his divinity, abases himself, takes
the form of a slave."[69] This is the meaning of St. John's insistence that the
Word was "in the beginning," as well as the meaning of the passage from the
Apocalypse in which St. John writes of "the Lamb slain from the foundation
of the world."[70] Our very existence comes at a supreme price, a price that only
God can pay.

Creation requires an eternal sacrifice, an eternal crucifixion of God, a sacrifice that can only be generated by divine love. In Weil's estimation, this eternal suffering and love not only is the source of our created world, but also is present throughout the universe, if we but learn to detect it.

> God created through love and for love. God did not create anything except love itself, and the means to love. . . . Because no other could do it, he himself went to the greatest possible distance, the infinite distance. This distance between God and God, this supreme tearing apart, this incomparable agony, this marvel of love, is the crucifixion. Nothing can be further from God than that which has been made accursed. This tearing apart, over which supreme love places the bond of supreme union, echoes perpetually across the universe in the depth of the silence, like two notes, separate yet blending into one, like a pure and heart-rending harmony. This is the Word of God. The whole creation is nothing but its vibration. When human music in its greatest purity pierces our soul, this is what we hear through it. When we have learnt to hear the silence, this is what we grasp, even more distinctly, through it.[71]

We shall see in the next chapter some of the practical implications for science of viewing creation's very existence as a suffering and dying motivated by love. For the present, it is enough to reflect on the possibility that the mathematical relationships and proportions which hold our universe together are signs of divine love, signs that it becomes our religious and moral duty to decipher and contemplate rather than merely manipulate. Much can be learned concerning the necessity and value of human suffering and affliction if one takes this paradigm to heart.

> But through the Creation, the Incarnation, and the Passion, there is also infinite distance. The interposed density of all space and all time sets an infinite distance between God and God. . . . Corresponding to the infinite virtue of unification belonging to this love there is the infinite separation over which it triumphs, which is the whole creation spread throughout the totality of space and time, consisting of mechanically brutal matter and interposed between

Christ and his Father. As for us men, our misery gives us the infinitely precious privilege of sharing in this distance placed between the Son and his Father. . . . This universe where we are living, and of which we form a minute particle, is the distance put by the divine Love between God and God. We are a point in this distance. Space, time, and the mechanism that governs matter are the distance.[72]

God and Humanity

As we have seen, the Greeks had a profound sense of the distance, the gulf between humanity and the divine. In Simone Weil's estimation, this sense of distance was not only important in the Greek understanding of humanity's place in reality, but also was prophetic of some of the foundational precepts of Christianity. For, at the heart of the Christian faith is the belief that whatever connection there may have been between God and humanity, this connection has been severed. Both the Greeks and Christians agree that this rupture can be healed only by a divine act of love. The following passage from "The Love of God and Affliction" powerfully portrays this separation:

> The infinity of space and time separates us from God. How can we seek for him? How can we go towards him? Even if we were to walk for endless centuries we should do no more than go round and round the world. Even in an airplane we could not do anything else. We are incapable of progressing vertically. We cannot take one step towards the heavens. God crosses the universe and comes to us.[73]

We saw in the previous section that the very existence of the created universe is dependent upon the continuing sacrifice of the Logos, the Son, as mediator between God and what is not God. In this section, we shall investigate the specific features of the divine mediation that makes loving contact between God and humanity possible. As Weil writes, "Insoluble contradictions have a supernatural solution. The solution of this one is the Passion."[74]

Although "it is true that there is a mysterious connexion between this distance and an original disobedience,"[75] Weil does not spend much time speculating concerning the cause of this rupture, what John Henry Newman called an "aboriginal calamity." The frequent misery of human existence, the apparent preponderance of evil over goodness—these are signposts of a "great gulf fixed" between God and humanity. Yet, given the considerations of the previous section, this rupture is to be expected, since the withdrawal of God is required for anything other than God to exist at all; there is no reason to believe that this does not apply fully to a very specific part of creation, human beings. Once we recognize that nothing "can be conceived more separately than man and God,"[76] we also recognize that "the only mediation between God and man is a being at once God and man."[77] God must become flesh (the Incarnation); the mediating activity of the God/human will be an instance in time of the divine mediating activity that makes creation possible. God, motivated by love, must die as a slave, a common criminal (the Passion). Only then can there be the perfection of friendship between the divine and the human. As we might expect, we begin our reflections on these profound mysteries with Simone Weil by turning once again to geometry and mathematics, as well as the Pythagorean claim that "friendship is an equality made of harmony," since "the Pythagorean definition of friendship, applied to God and to man, makes mediation appear as being essentially love, and love as being essentially the mediator."[78]

One of Weil's favorite words to describe the mediation that must take place in order for God and humanity to become friends is "assimilation." It is from the Latin verb *assimilare,* which itself is a translation of the Greek verb *homoio,* meaning "to make like," and the noun *homoioma,* a "becoming like, resemblance." If there is to be assimilation between God and humanity, Weil suggests, we must have faith in the impossible, a mediation between entirely incommensurable properties. "Assimilation provides the one and only contact with God, and faith in the reality of this contact implies faith in the possibility of mediation."[79] The word translated as "assimilation," however, is also a mathematical word, the one used in the passage from the *Epinomis* considered briefly in a previous chapter.

> Geometry . . . is the assimilation of numbers which do not by nature resemble one another, an assimilation made manifest when referred to the properties of plane figures. That this marvel is not human but divine is clear to anyone who can think. . . . What is supernatural

and miraculous for those who can contemplate and think is that . . .
all of nature is marked by the form and the essence of the relationship of opposites, according to each proportion.[80]

As we have seen, "the assimilation between two numbers meant the discovery of a proportional mean,"[81] a mean that can be established even when the
two numbers in question are incommensurable. Hence "the assimilation of
man to God meant the discovery of a mediation,"[82] mediation that, Weil
insists, is perfectly mirrored by the process of mediation between incommensurable magnitudes. Recall Weil's belief that the point of mathematics is
not mathematics *per se,* but the search for bridges between the divine and
human. In a profound sense, the Greek search for mean proportionals was
prophetic of the greatest mean proportional of all, the Christ.

> The text in the *Epinomis*: . . . the assimilation of numbers which
> are not by their nature similar, a marvel (or miracle) not of human
> but divine agency, which the genus of plane figures was destined to
> reveal—doesn't this show that the Greeks saw in the geometrical
> mean an image of the Incarnation? And the proportion with which
> they were obsessed, which they sought everywhere, in the whole of
> their science, in the whole of their art, was it not for this reason that
> they were, in fact, so obsessed with it?[83]

What exactly does "assimilation" mean in the mathematical sense?
Recall Weil's discussion of mediation between numbers. Some numbers
(squares) have a natural mediation with unity through the number of which
they are a square. For instance, the number 9 is naturally mediated with 1
via its square root 3 in the relationship $1/3 = 3/9$. It is in reference to such
numbers that the Pythagoreans cryptically claimed that "justice is a number
to the second power"; these numbers that are naturally linked to unity by a
mean proportional that is a whole number "are an image of perfect justice."[84]
We have also seen, however, that such special numbers are rare. More often,
there is no such natural mediating link between numbers; in many cases,
numbers are incommensurable with each other. Yet we also know that discoveries in geometry revealed how all numbers, even those that are incommensurable, can be mediated by using the properties of right triangles and

circles. Eudoxus showed how, by expanding the concept of number to include comparative relationships, the mediation of incommensurable magnitudes could be expressed arithmetically.

What has any of this to do with mediation between God and humanity? Weil provides us with a suggestion that opens the door to an almost direct transfer of the truths contained in the mediation of incommensurables in mathematics to the realm of the transcendent.

> If one considers whole numbers, one sees they are of two sorts; those that are linked to unity by a mean proportional, such as 4, 9, 16 on one hand, and on the other hand all the others. If the first are an image of perfect justice, as the Pythagoreans say, we resemble the others, we who are in sin. Was it by force of so intense a search for a mediation of these wretched numbers that the Greeks discovered geometry?[85]

Weil wishes us to consider the incommensurability of many numbers with each other as a picture of the incommensurability between God and humanity. She suggests in her Marseilles notebooks that "the gods are the numbers which are square—(the angels)—Men are the other numbers. Moreover, something has necessarily got to intervene to bring about mediation."[86] We know that in mathematics, finding mean proportionals between incommensurables required thinking outside normal parameters, required an invasion from "outside." Precisely the same sort of activity outside normal limits is required to mediate between God and humanity.

> Justice is that between which and God there is naturally mediation. On the other hand, there is not naturally mediation between sinners and god (they are "numbers not naturally similar"), just as there is not between unity and numbers other than square. But in the same way that geometry, through the destiny of plane figures, supplies a miraculous mediation for these numbers, so there is a miraculous operation, contrary to nature, which establishes a mediation between criminal humanity and God ("assimilates to one another numbers not naturally similar").[87]

What evidence is there that we should understand the discovery of mean proportionals between incommensurables as a sign of the mediating activity of Christ to come? Weil draws our attention to the various New Testament passages in which Christ refers to himself as a mediator, as well as the passages in the Gospel of St. John considered in the previous section.

> Geometry is the science of the search for proportional means (similar, in Greek, signifies proportional) by way of the incommensurable proportion . . . it proceeds from a supernatural revelation. If we compare this with the passages in which Plato describes the mediation between God and man by the image of the proportional mean (a/b = b/c) and with the numerous Gospel passages in which Christ uses the same image for his own function as mediator, then the invention of geometry in Greece in the sixth century B.C. appears to be in the strictest sense of the word a prophecy.[88]

Furthermore, Weil argues, Christ himself was fully aware that he, in his dual role as God and human, was the fulfillment of this prophecy.

> The allusion is evident. Just as the Christ recognized Himself as Isaiah's man of sorrows, and the Messiah of all the prophets of Israel, He recognized Himself also as being that mean proportional of which the Greeks had for centuries been thinking so intensely.[89]

It remains now to investigate the dynamics of this divine mediation, guided by what we know concerning both geometry and the mediation between Creator and created discussed in the previous section.

In an essay entitled "The 'Republic,'" Weil provides a helpful illustration of the mathematical meaning of "assimilation" as used by Plato.

> When Plato speaks of assimilation . . . the word assimilation is used in the sense which we give it today, it is a question of resemblance. Only the meaning is more rigorous, it is of a resemblance such as exists between two geographical maps of two different scales, wherein the distances are different but the relationships are identical. For

the word assimilation in Greek, and especially for such a Pythagorean as Plato, is a geometrical term which refers to the identity of
relationships, to proportion. When Plato speaks of assimilation in
God, it is no longer a question of resemblance, for no resemblance is
possible, but one of proportion. No proportion is possible between
men and God except by mediation. The divine model, the perfectly
just man, is the mediator between just men and God.[90]

Weil's argument is that this mathematical understanding of assimilation is
necessary when considering Christ's mediating activity between God and
humanity. As she writes in her Marseilles notebooks, "the assimilation to
God can only be brought about by a proportional mean. . . . The mean
(*metaxu*) is the perfectly just man: God/God-man = God-man/man."[91] Although divine mediation creates a mean proportional between God and
humanity, establishing a friendship between them, there is no sense of
humanity being raised to direct equality with God.[92] Rather, what is created
is the harmony between incommensurable quantities that, in themselves,
still remain incommensurable.

> Harmony is the principle of this sort of equality, harmony which is
> the bond between the contraries, the proportional mean, the Christ.
> It is not directly between God and man that there is something
> analogous to a bond of equality, it is between two relationships.[93]

This sort of harmony is illustrated by the simple mathematical propor-
tion $1/\sqrt{3} = \sqrt{3}/3$. In this equation, $\sqrt{3}$ is the mean proportional between 1
and 3, but $\sqrt{3}$ is an irrational number. Still, there is an equality between the
two relationships expressed here, even though something outside the realm
of rational numbers is required to establish the equality. Similarly, Christ as
the mediator between God and humanity establishes an equality of relationship, in which the parties are bound together in equality of relationship: the
Father is to the Son as the Son is to humanity. Only a mediator which par-
takes of both extremes of the relationship (God-human) can achieve the
mediation. Once again, Weil finds that geometry is prophetic of Christ's
mediating activity.

When Plato . . . speaks of geometric equality, this expression is doubtless exactly equivalent to that of harmonious equality employed by Pythagoras . . . rigorously defined by the equality between two ratios having a common term, of the type a/b = b/c. For the adjective geometric, in such terms as geometric mean and geometric progression, indicates proportion. The phrases from St. John cited above have so clearly and insistently the aspect of an algebraic equation that this is manifestly what is meant and what allusion is made to.[94]

In addition to a model of divine mediation, we can learn much concerning the very nature of Christ's mediating activity between God and humanity by pushing the analogy with incommensurable magnitudes a bit further. In the previous section we found that the mediating activity of the Logos between Creator and creation requires a continuing supreme sacrifice; in the present case, the mediating activity of Christ requires a time-bound sacrifice paralleling that of the Logos. Weil draws on Plato's discussion of the just person in the *Republic* to establish what the true nature of a mediator between a perfectly just God and sinful humanity must be. In order for there to be mediation, a perfectly just human being must exist in reality, not just in the abstract. "If the model [the perfectly just man] is to be real, he must have an earthly existence at a certain point in space, and at a certain moment of time. . . . If the ideal cannot have this existence, it is nothing but an abstraction."[95] In other words, this person must become flesh, must become incarnate. Furthermore, as Plato argued, perfect justice and mediation will not be recognized as such; just as Prometheus suffered because he sought to mediate between the gods and humanity, so the divine mediator's activity will be rooted in suffering, punishment, abandonment, and death. "This essentially and irreducibly penal character of redemptive suffering is what the Greeks understood very well."[96] It is also at the very heart of the Christian faith.

Throughout the final years of her life, Simone Weil focused on the importance of the Passion and the Cross far more often than that of the Resurrection when considering the truths of the Christian faith. Although there are many reasons for this emphasis, it is already at least partially apparent why she believed the Cross to be more important than the Resurrection. The Cross is the point of intersection between God and humanity, the point at which Christ achieved mediation. "The death on the Cross is something more divine than the Resurrection, it is the point where Christ's divinity is concentrated."[97] It is also the point at which the God-man, the Christ, was most distant from and abandoned by God: "My God, my God, why hast thou forsaken me?" It is this separation, motivated entirely by divine love, that achieves mediation between God and creation, God and humanity. "Supreme mediation, the harmony between Christ's 'Why?' (ceaselessly repeated by every soul in affliction) and the silence of the Father. The universe (ourselves included) is the vibration of this harmony."[98]

Another reason that Weil focuses on the Cross rather than the Resurrection is to counter our common tendency to emphasize the Resurrection, thus failing to recognize the depth of the agony and suffering required for Christ's mediation. She does not deny the Resurrection; rather, she asks us not to let our joy at the risen Christ diminish our understanding of the price required for us to be made the friends of God.

> During the days when Christ was completely stripped of all appearance of justice, even his friends themselves were no longer wholly conscious of his being perfectly righteous. Otherwise could they have slept while he suffered, could they have fled, have denied him? After the Resurrection the infamous character of his ordeal was effaced by glory, and today, across twenty centuries of adoration, the degradation which is the very essence of the Passion is hardly felt by us. . . . We no longer imagine the dying Christ as a common criminal.[99]

Even St. Paul argued that the focal point of the Christian faith is the Resurrection: "If Christ be not raised, your faith is vain."[100] Weil, however, asks us not to forget that, more foundationally, the Christian faith is vain if there is no Cross, no suffering and death of the divine mediator between God and humanity. Only when we see, as did the penitent thief, that the criminal hanging on a cross, rejected and despised by all, is the perfectly just God-man paying the ultimate sacrifice to achieve mediation between God and humanity will we begin to truly experience the mystery of the Christian faith and the foreshadowings of it found in the geometry and mathematics of the Greeks.

Today the glorious Christ veils from us the Christ who was made a malediction; and thus we are in danger of adoring in his name the appearance, and not the reality, of justice. In short, only the penitent thief has seen justice as Plato conceived it, naked and perfect, veiled beneath the appearance of a criminal.[101]

Clearly such an understanding of how geometry and mathematics point toward something greater and more profound than the intellect can grasp has important implications for the interactions between human beings themselves as well as for how human beings understand the world around them. These will be the issues investigated in the next chapter. Here, it is important to remind ourselves that, for Weil, true Christianity is the perfection of the Greek yearning for mediation and bridges between the divine and humanity, a yearning strikingly worked out in their investigations into geometry and mathematics. Even now, centuries later, a simple struggle with a geometry problem can orient us toward the transcendent.

The solution of a geometry problem does not in itself constitute a precious gift, but the same law applies to it because it is the image of something precious. Being a little fragment of particular truth, it is a pure image of the unique, eternal, and living Truth, the very Truth that once in a human voice declared: 'I am the Truth.'[102]

CHAPTER SIX

"Our Country Is the Cross"

By loving our neighbor we imitate the divine love which created us and
all our fellows. By loving the order of the world we imitate the divine love
which created this universe of which we are a part.

—Simone Weil, "Forms of the Implicit Love of God," *WG,* 158

Although many of Simone Weil's writings, including the majority of
those that have been the focus of this study thus far, point us in directions
other than the practical, daily struggle of human existence, she was intensely
interested in the question of how a human being should live his or her life in
this world. Undoubtedly the transcendent, the "supernatural," is a crucial
theme in her thought, especially in her later writings; ultimately, however,
she insists that "the object of my search is not the supernatural, but this
world,"[1] calling for us to learn to "love the country of here below. It is real; it
offers resistance to love. It is this country that God has given us to love. He
has willed that it should be difficult yet possible to love it."[2] In this chapter, we will investigate how the Pythagorean dictum that "friendship is an
equality made of harmony" can shed light on human beings, both in their
interactions with each other and their relationships as inhabitants of the
earth with other things in this world. Just as in the three relationships discussed in the previous chapter (the Trinity, Creator/created, God/humanity),

we shall find that the key to understanding human relationships and the proper orientation of the human being toward the rest of creation will be the mediation of opposites. This, in turn, provides guidance both in the realm of moral behavior and scientific enquiry.

We have seen that the very existence of creation itself, including human existence, is due to unending divine love and grace. The tension between union and separation, the pain and joy of existence itself, is imprinted in the very fabric of human nature. We are incarnated contradictions, "delivered over to appearances and to sorrows and desires, and yet destined for something other . . . infinitely different from God and yet obliged to be perfect like [our] heavenly Father."[3] Yet within our very nature, Weil argues, we find signs and clues that, if read properly, can orient us toward each other, toward the world, and ultimately toward God.

> There are three mysteries in human life of which all human beings, even the most mediocre, have more or less knowledge. One is beauty. Another is the work of pure intelligence applied to the contemplation of theoretic necessity in the understanding of the world, and the incarnation of these purely theoretical conceptions in technique and in work. The last are those flashes of justice, of compassion, of gratitude which rise up sometimes in human relationships in the midst of harshness and metallic coldness. Here are three supernatural mysteries constantly present right in human nature. These are three openings which give direct access to the central door which is the Christ.[4]

In this chapter we will consider these "three mysteries" in reverse order, investigating first the "flashes" of justice and compassion that can be found in human relationships, then considering how the proper investigation of the world around us reveals the sources of beauty and order that must necessarily inform any true science. The overarching theme will be the various ways in which understanding ourselves and our world can orient us toward God.

Weil is acutely aware of the pain and difficulty of the human condition, features of reality that, far from revealing divine love, cause many to despair of finding meaning in the world and in life.

Human life is impossible . . . the good is impossible. . . . Our life is impossibility, absurdity. Everything we want contradicts the conditions or the consequences attached to it, every affirmation we put forward involves a contradictory affirmation, all our feelings are mixed up with their opposites.[5]

Weil's argument will be that the difficulty of human existence is a reflection both of the contradictory elements in human nature and of the continuing sacrifice required for us to exist at all. Her aim is to convince us that all relationships in the world, whether they include human beings or not, are ultimately reflective of divine love and providence.

Every relation between two or several created things—whether thinking beings or matter—is one of God's thoughts. We ought to desire a revelation of the thought of God corresponding to each relation with our fellow men or with the material objects with which we are involved. . . . The creation is a tissue of God's particular thoughts. Each of us is a knot of those thoughts. . . . We have to become such that every one of our thoughts—that is to say every relation of our soul with everything connected with it by any relation in the past, present, or future—shall coincide with a particular thought of God's.[6]

This is a difficult challenge, but Weil reminds us that we are sorely mistaken to expect that human existence will be anything but frequently difficult and painful. Our very existence requires, our very nature reflects, the ultimate sacrifice.

The Trinity and the Cross are the two poles of Christianity, the two essential truths: the first, perfect joy; the second, perfect affliction. It is necessary to know both the one and the other and their mysterious unity, but the human condition in this world places us infinitely far from the Trinity, at the very foot of the Cross. Our country is the Cross.[7]

HUMAN AND HUMAN

Although the impression is sometimes given that Simone Weil was neither particularly concerned about nor especially good at interpersonal relationships, this impression is countered by her apparent long-standing interest
in the concept of friendship. As early as her prewar notebooks, she struggled
with understanding the nature of friendship as well as how to separate it
from the sentimental trappings that often accompany relationships going by
the name of friendship. Coming to an understanding of this most profound
of human relationships seemed, to her, to be a necessary component of the
maturation process.

> Friendship is not be sought for, dreamed about, longed for, but exer-
> cised (it is a virtue). . . . There is certainly no need to ignore the
> inspiring virtue of friendship. But what ought to be strictly prohib-
> ited is dreaming of sentimental delights. . . . At the age of 25 it is
> high time for a radical break with adolescence. . . .[8]

Although Weil's conception of friendship deepened in insight and complexity
during the decade or so after this notebook entry, she never changed her profound conviction that "friendship, like beauty, is a miracle. And the miracle
consists simply in the fact that it *exists*."[9]

Athens and Melos

The most familiar use of the term "friendship" specifies a special, close
relationship of love and affection between two human beings. In the context
of our present study, however, we are most interested in the broader concept
defined by the Pythagorean definition of friendship. If we consider this definition closely, it is not immediately apparent how it can apply to relationships
between human beings. If "friendship is an equality made of harmony," and
if harmony has to do with mediation between opposites, then in what sense
can this apply to friendship on the human level?

> The same [Pythagorean] definition applies also to friendship be-
> tween men, although there is more difficulty in this since as

> Philolaus said: "Things of the same species, of the same root, and of
> the same station, have no need of harmony." It is significant that the
> Pythagoreans should have chosen a definition of friendship which
> does not apply to relationships between men except in the last place.
> Friendship is first friendship in God between the divine Persons.
> From this follows friendship between God and man. In the last
> place only is it friendship between two men or more.[10]

The opposites involved in the relationships of the Trinity, Creator/created,
and God/humanity are clearly in need of supernatural mediation, as our discussion in the previous chapter showed. The present case, however, appears
to be different. In what sense are human beings, identical in species and
similar in countless ways, in need of the harmony that is a union of opposites? We begin with an investigation of the natural relationships between
human beings.

Weil's answer to the above question, an answer that ultimately causes
her to conclude that human individuals are by nature contraries so opposed
that only divine mediation can cause true friendship between them to occur,
begins with the observation that the Pythagorean notion of friendship applies to general human interaction because "although they [humans] are in
fact of the same species, of the same root, of the same rank, they are not so
in their thought."[11] Each human individual is separated from all other individuals by a matter of first-person perspective. In human relationships, the
contrariety that requires mediation, proportionality, and harmony is the opposition between the "I," whether individual or collective, and every other
human being. In each person, "I" is "the center of the world"—for the most
part, "everyone disposes of others as he disposes of inert things, either in
fact, if he has the power, or in thought."[12] Human relationships, in natural
terms, are about power, each party seeking to preserve autonomy at the expense of the autonomy of the other party.

According to Weil, the key factor in meditating between the "I" and the
"other" is mutual *consent;* she notes in her essay "Are We Struggling for Justice" that "the Greeks defined justice admirably as mutual consent," and
quotes the passage from the *Symposium* in which Plato writes that "where
there is agreement by mutual consent there is justice, say the laws of the
royal city."[13] There are instances in which human beings need each others'

consent and in which they are, for all practical purposes, "equal" in terms of need or power. As she writes in "Forms of the Implicit Love of God," "when two human beings have to settle something and neither has the power to impose anything on the other, they have to come to an understanding. Then justice is consulted, for justice alone has the power to make two wills coincide."[14] In such cases, justice can be established between them without sacrificing the "I"—"justice then occurs as a natural phenomenon."[15] Such justice is valuable and should be sought as often as possible, but such "natural" justice "does not constitute harmony, and it is a justice without friendship."[16] The cryptic Pythagorean notion that "justice is a number to the second power" illustrates such a situation. Just as 9 has a special bond with unity through 3, so justice can occasionally be established on a natural basis.

Weil's position, however, is that such natural justice is exceptionally rare. More often than not, the natural relations between human beings do not reveal equal need for consent, just as more often than not, the numbers needing mediation with unity are irrational or "wretched." As she writes in her New York notebooks, "in nature, either I am at the center (perspective) or another person is, who dominates me by brute force, and the rest is simply pieces of the universe, apart from the exceptional case of natural justice."[17] Weil provides a striking illustration of what she means in "Forms of the Implicit Love of God."

> Justice alone has the power to make two wills coincide. . . . But when there is a strong and a weak there is no need to unite their wills. There is only one will, that of the strong. The weak obeys. Everything happens just as it does when a man is handling matter. There are not two wills to be made to coincide. The man wills and the matter submits. The weak are like things. There is no difference between throwing a stone to get rid of a troublesome dog and saying to a slave: "Chase that dog away."[18]

If it is truly the case that "men are unequal in all their relations with the things of this world, without exception,"[19] and are thus unequal with each other, humanly speaking, then the equality required for friendship between human beings in the Pythagorean sense can only be established through supernatural mediation. "When one applies the formula 'Friendship is an equality made of harmony' to men, harmony has the meaning of the unity of contraries. The contraries are myself and the other, contraries so distant that they have their unity only in God."[20] In human relationships, "friendship is identical with supernatural justice."[21]

One of Weil's favorite examples of the natural relations between human beings, discussed in several of her writings, is the interaction between the Athenians and Melians from Thucydides' *The Peloponnesian War.*[22] The Melians, traditional allies of Athens' enemy Sparta, were forced by the Athenians to choose between becoming the allies of Athens or having their community destroyed. In response to the Melian call for justice and fair treatment, the Athenians, who are in a position of power, reply:

> The human mind being made as it is, the justice of a matter is examined only if there is an equal necessity on both sides. Contrarily, if one is strong and the other weak, what is possible is accomplished by the first and accepted by the second. . . . Of the gods, we believe, as of man, we certainly know, that it is a necessary law of their nature to rule wherever they can.[23]

The seemingly amoral position of the Athenians, who ultimately raze Melos to the ground, put all the men of Melos to death, and sell the Melian women and children into slavery, is in their estimation simply hard realism. We are all familiar with such realism, a position given expression by Friedrich Nietzsche in the following comments on the Athenian/Melian conflict from *The Will to Power:*

> Do you suppose perchance that these little Greek free cities, which from rage and envy would have liked to devour each other, were guided by philanthropic and righteous principles? Does one reproach Thucydides for the words he put into the mouths of the Athenian ambassadors when they negotiated with the Melians on the question of destruction or submission? Only complete Tartuffes could possibly have talked of virtue in the midst of this terrible tension— or men living apart, hermits, refugees, and emigrants from reality— people who negated in order to be able to live themselves.[24]

Events more than two thousand years later in Bosnia and Kosovo, just a few hundred miles from Melos, confirm the continuing power of such hard realism. Interestingly, Weil argues that on a natural basis the Athenians are entirely correct. Just as there is no "natural" mediation between numbers that are incommensurable, so there is no natural basis for consent between human beings when there exists neither equal need nor equal power. There is no natural reason for the Athenians to seek the consent of the Melians— "it would be absurd and mad for anyone at all to impose upon himself the necessity of seeking consent where there is no power of refusal."[25]

The mediation of incommensurables on the human level requires, as it were, an "abdication" of the "I," an abdication modeled by the Christian doctrine of the Incarnation, according to which Christ "assumed the state of slavery . . . He humbled himself to the point of being made obedient unto death."[26] Here is an example of what is truly of most value, the Divine itself, choosing voluntarily to become equal to, even lower than, what is truly lower than itself. Weil observes, however, that such an argument would hardly have been convincing to the Athenians. "These words could have been an answer to the Athenian murderers of Melos. They would have really made them laugh. And rightly so. They are absurd. They are mad."[27] Their absurdity and madness arises because the required mediation is impossible without an "invasion" from a domain other than the natural.

The supernatural element needed to establish harmony and equality between unequal human beings places such relationships within the scope of the Pythagorean definition of friendship, which is the same as supernatural justice. "The supernatural virtue of justice consists of behaving exactly as though there were equality when one is the stronger in an unequal relationship."[28] Weil believes that the construction of the mean proportional between incommensurates provides us with a direct picture of the only possible source of the mediation needed between human beings that are "incommensurate." Just as the mediation of incommensurable magnitudes requires a mediation from an outside domain, so the mediation between human incommensurates, whether the strong and the weak or the "I" and the "other" must come from a domain or source that is "outside." The mediator must be truly "other," or "divine." This is why Weil insists that ultimately "human consent is a sacred thing."[29] "Supernatural justice, supernatural friendship or love, are found to be implicit in all human relationships where, without there being an equality of force or of need, there is a search for mutual consent."[30] It is precisely this supernatural source of mediation between contraries that is at the heart of the Christian faith. To the Athenian claim that "Of the gods, we believe, as of man, we certainly know, that it is a necessary law of their nature to rule wherever they can," Weil responds that "the Christian faith is nothing but the cry affirming the contrary."[31]

Something to Love

What could cause a person to treat another person as an equal, when there is no "earthly" reason to do so? One must seek for something in the other person that transcends the various contingencies that establish hierarchies and inequality. Weil, writing toward the end of World War II, observes that "mankind has become mad from want of love."[32] Such madness can be addressed only by another sort of madness, a love that "once it has seized a human being, completely transforms the modalities of action and thought."[33] Such love, as any love, requires an object.

> One needs above all something to love . . . not through hating its opposite, but in itself. . . . Something to love not for its glory, its prestige, its glitter, its conquests, its radiance, its future prospects, but for itself. . . . What we need is something people can love naturally from the depths of their hearts.[34]

In "Draft for a Statement of Human Obligations," Weil describes the metaphysical framework within which one can locate the true source of human equality.

> There is a reality outside the world, that is to say, outside space and time, outside man's mental universe, outside any sphere whatsoever that is accessible to human faculties. Corresponding to this reality, at the center of the human heart, is the longing for an absolute good, a longing which is always there and is never appeased by any object in this world. Just as the reality of this world is the sole foundation of facts, so that other reality is the sole foundation of the good. That reality is the unique source of all the good that can exist in this

world: that is to say, all beauty, all truth, all justice, all legitimacy, all order, and all human behavior that is mindful of obligations. Those minds whose attention and love are turned towards that reality are the sole intermediary through which good can descend from there and come among men. Although it is beyond the reach of any human faculties, man has the power of turning his attention and love towards it.[35]

The task of the person seeking to establish equality and friendship with another person who is naturally unequal is to turn his or her attention toward this reality, with the conviction that such attention will not go unrewarded. "To anyone who does actually consent to directing his attention and love beyond the world, towards the reality that exists outside the reach of all human faculties, it is given to succeed in doing so."[36]

Weil locates the true basis of human equality in the human capacity to seek for and believe in the efficacy of what is greater than oneself.

The combination of these two facts—the longing in the depth of the heart for absolute good, and the power, though only latent, of directing attention and love to a reality beyond the world and of receiving good from it—constitutes a link which attaches every man without exception to that other reality. Whoever recognizes that reality recognizes also that link. Because of it, he holds every human being without any exception as something sacred to which he is bound to show respect. This is the only possible motive for universal respect towards all human beings.[37]

The object of the love required to establish the equality of all persons despite their contingent differences is precisely this capacity, equal in all persons, to attend to what transcends contingency.

The only thing that is identical in all men is the presence of a link with the reality outside the world. All human beings are absolutely identical insofar as they can be thought of as consisting of a center, which is an unquenchable desire for good, surrounded by an accretion of physical and bodily matter.[38]

This is why Weil reminds herself "to accord value in myself only to what is transcendent";[39] only in the context of what is transcendent can the natural human tendency to value oneself more highly than others be countered by one's attention to what is truly of value in all persons. This, in turn, makes justice and equality possible.

> Anyone whose attention and love are really directed towards the reality outside the world recognizes at the same time that he is bound, both in public and private life, by the single and permanent obligation to remedy, according to his responsibilities and to the extent of his power, all the privations of soul and body which are liable to destroy or damage the earthly life of any human being whatsoever.[40]

Weil most often expands on the human ability to treat unequals as if they were equal within the context of those persons who are "afflicted," crushed by the contingencies and circumstances of life to the point of being more like a thing than a person.

> Christ taught us that the supernatural love of our neighbor is the exchange of compassion and gratitude which happens in a flash between two beings, one possessing and the other deprived of human personality. One of the two is only a little piece of flesh, naked, inert, and bleeding beside a ditch; he is nameless; no one knows anything about him. Those who pass by this thing scarcely notice it, and a few minutes afterward do not even know that they saw it.[41]

Not surprisingly, Weil frequently uses the story of the Good Samaritan to illustrate the supernatural dynamic of human justice and friendship. The miracle expressed in this story is that the broken and dying man in the ditch can only be made a person again by a human being turning attention toward him and choosing to impart personhood where it is absent.

> Creative attention means really giving our attention to what does not exist. Humanity does not exist in the anonymous flesh lying

inert by the roadside. The Samaritan who stops and looks gives his attention all the same to this absent humanity, and the actions which follow prove that it is a question of real attention.[42]

Although the Good Samaritan and the man robbed by thieves were strangers, the Good Samaritan saw the injured man as far more than an arbitrary example of humanity. It was this particular man in this particular circumstance with these particular needs that the Samaritan attended to. In doing so, the Samaritan restored the man's humanity and self-respect.[43]

> The preservation of true self-respect in affliction is something supernatural. . . . The afflicted man and his benefactor, between whom diversity of fortune places an infinite distance, are united. . . . There is friendship between them in the sense of the Pythagoreans, miraculous harmony.[44]

Weil argues that such a mediation between unequals is best understood as a reflection of the divine act of mediation that establishes the bond between Creator and creation, as well as that between God and humanity. "He who treats as equals those who are far below him in strength really makes them a gift of the quality of human beings, of which fate had deprived them. As far as it is possible for a creature, he reproduces the original generosity of the Creator with regard to them."[45] Only a belief in the possibility of divine mediation, as modeled by the mediation between incommensurables in mathematics, can provide a framework within which the hard realism of the Athenians can be addressed.

> The Athenians of Thucydides thought that divinity, like humanity in its natural state, always carried its power of commanding to the extreme limit of possibility. The true God is the God we think of as almighty, but as not exercising his power everywhere, for he is found only in the heavens or in secret here below. Those of the Athenians who massacred the inhabitants of Melos had no longer any idea of such a God. The first proof that they were in the wrong lies in the fact that, contrary to their assertion, it happens, although extremely

rarely, that a man will forbear out of pure generosity to command
where he has the power to do so. That which is possible for man is
possible also for God.[46]

Once again, the unity of opposites, this time between human beings, requires
a supernatural spark of divinity. "It is impossible for two human beings to be
one while scrupulously respecting the distance that separates them, unless
God is present in each of them."[47]

THINGS AND THINGS

With the fifth and final relationship expressive of the "marvellous
meanings" that Weil finds embedded in the Pythagorean definition of friendship, we encounter the relationship between human beings and things in the
material world.

> Besides that of the Trinity, of the Incarnation, of the charity between
> God and man, and of the charity between men, there is a fifth form
> of harmony, which concerns things. This also surrounds man in so
> far as a man is a thing, that is to say, the whole man, body and soul,
> except the faculty of free consent.[48]

Within the broad context of our present study, this relationship may be the
most important of all to understand properly, since the relationship of humanity to the material world is of crucial importance to a proper framework for
science, the study of the world around us. We shall find that Weil's investigation of this relationship will bring us full circle to her early thinking on the
manner in which human beings encounter and "create" their world cogni-
tively, as discussed in chapter 1. In her mature thinking, these early investigations are enriched by her continuing immersion in mathematics as well as
by her increasing attraction to transcendent paradigms. In addition, we now
return directly to the mathematical and geometrical order of the world, which
now, as necessity, reveals the heart of God. In the investigation of the world,
properly framed, we will once again encounter the divine.

When we are alone in the heart of nature and disposed to give it attention, something inclines us to love what surrounds us, which, however, consists only of brute matter, inert, dumb and deaf. And beauty touches us all the more keenly where necessity appears in a most manifest manner, for example in the folds that gravity has impressed upon the mountains, on the waves of the sea, or on the course of the stars. . . . This cluster of marvels is perfected by the presence, in the necessary connections which compose the universal order, of divine verities symbolically expressed. Herein is the marvel of marvels, and as it were, the secret signature of the artist.[49]

The key themes in our investigations of the interactions between human beings and the world will include necessity, consent, obedience, and beauty. Weil finds parallels between the relationships amongst human beings, investigated in the previous section, and the equilibrium that must be identified and sought if human beings are to occupy their proper place in the order of the material world. In both cases, the proper autonomy of the things in question must be identified according to proper boundaries, then respected. The order of the world is as it should be; respecting this order requires the conviction that this order is also good and reflective of divine love.

This necessity constitutes an order whereby each thing, being in its place, permits all other things to exist. The maintenance of boundaries constitutes for material things the equivalent of what the consent to the existence of others is for the human spirit, that is to say, charity for one's neighbor.[50]

And, just as true charity for one's neighbor requires the removal of the illusion that one's self is the center of the universe, so the proper love and respect for the world requires that the science centered on power and control that has characterized the West over the past several centuries be transformed by recognition of the proper center for all human activity.

To empty ourselves of our false divinity, to deny ourselves, to give up being the center of the world in imagination, to discern that all points in the world are equally centers and that the true center is outside the world, this is to consent to the rule of mechanical necessity in matter and of free choice at the center of each soul. Such consent is love. The face of this love, which is turned toward thinking persons, is the love of our neighbor; the face turned toward matter is love of the order of the world, or love of the beauty of the world which is the same thing.[51]

Necessity

There is perhaps no concept in Simone Weil's writings that is of more importance, yet more complex and difficult, than necessity.[52] Its importance to our present considerations is clearly expressed in the following passage from "The Love of God and Affliction":

> The idea of necessity as the material common to art, science, and every kind of labor is the door by which Christianity can enter profane life and permeate the whole of it. For the Cross is necessity itself brought into contact with the lowest and the highest part of us; with our physical sensibility by its evocation of physical pain and with supernatural love by the presence of God. It thus involves the whole range of contacts with necessity which are possible for the intermediate parts of our being.[53]

Necessity, according to Weil, is the word that best expresses the mathematical laws of variation and proportion that have been the frequent focus of our considerations. "The word *Logos*, borrowed from the Greek Stoics who had received it from Heraclitus, has several meanings, but the principal one is that quantitative law of variation which constitutes necessity."[54] Hence we can expect that a number of the various threads of our discussion to this point will be brought together under the concept of necessity.

Weil orients us to the special nature of the relationship between humanity and the world around us by noting that this relationship is somewhat different than those relationships previously considered that satisfied the Pythagorean definition of friendship.

This fifth form of harmony, not being concerned with persons, does not constitute a friendship. The contraries to which it is related are the principle which limits, and that which receives limits from outside, which is to say: God and inert matter as such. The intermediary is the limit, the network of limits, which hold all things in a single order.[55]

On the next page, she identifies this network of limits, the mathematical and geometrical network imposed by God on matter, as necessity. "Necessity always appears to us as an ensemble of laws of variation, determined by fixed relationships and invariants."[56] We have already seen these laws of variation expressed mathematically as the relationship, for instance, between an object and its shadow. In a larger sense, the order, regularity, and structure of the world itself is due entirely to necessity, what Weil identifies as "the intermediary between matter and God."[57]

The rainbow's beautiful semicircle is the testimony that the phenomena of this world, however terrifying they may be, are all subject to a limit. . . . Like the oscillations of the waves, the whole succession of events here below, made up, as they are, of variations in balance mutually compensated—births and destructions, waxings and wanings—render one keenly alive to the invisible presence of plexus of limits without substance and yet harder than any diamond.[58]

In short, Weil says, "necessity is the obedience of matter to God."[59]

It is important to recognize that the structure of the universe, as we perceive it, tells us nothing about matter, the unknown "something" that is brought within limits by divinely imposed necessity. The impossibility of "getting at" the reality of matter itself is something the Greeks (and subsequent thinkers) were entirely aware of.

For us the reality of the universe is necessity whose structure is that of the gnomon, supported by something. Necessity must have a support, for by itself it is essentially conditional. Without a basis, it is but abstraction; upon a basis it constitutes the reality of creation itself. Of that basis we cannot have the least conception. However the Greeks had a word (*apeiron*), which means at once unlimited and indeterminate. For us, matter is simply what is subjected to necessity. We know nothing else about it.[60]

What we do know, however, and have a special relationship to on many levels, is necessity. "Reality for the human mind is contact with necessity."[61]

Our ability to know anything at all is rooted in our ability to recognize in the various phenomena of our existence the regularities, limits, and functional relationships that constitute necessity.

> The knowledge of sensible phenomena is uniquely the recognition in these phenomena of something analogous to that purely conditional necessity. The same is true for psychological and social phenomena. One knows them so far as one recognizes in them, concretely and precisely at each occasion, the presence of a necessity analogous to mathematical necessity. This is why the Pythagoreans said that we know nothing but number.[62]

Abstracted from specific phenomena, the regularities of the world "appear to the mind as a network of relations which are immaterial and without force. Necessity can only be perfectly conceived so long as such relations appear as absolutely immaterial."[63] And, Weil argues, it is such a conception that is required in order for true mathematics and science to be possible.

> To think of necessity in a way that is pure, it must be detached from the matter which supports it and conceived as a fabric of conditions knotted one with the others. Necessity, both pure and conditional, is the true object of mathematics and of certain operations of thought which are analogous to mathematics; which are as theoretical, pure, and rigorous as mathematics but which are not given a name because they are not discerned.[64]

Once we understand that necessity is an ensemble of immaterial limiting and ordering relationships rather than a "thing," we have taken a significant step in the direction of not only understanding the nature of the mathematical structure of our world, but also of the important contribution that we, as human knowers, make to this very structure. Recalling that Weil argued in texts as early as "Science and Perception in Descartes" that the construction of the world for the human knower is an activity involving both knower and world, we see now that it is the human mind itself that illuminates the very relationships that define reality.

> It is not in our power to modify the sum of the squares of the sides in the right-angled triangle, but there is no sum if the mind does not work it out by conceiving the demonstration. Already in the domain of whole numbers, one and one can remain side by side throughout perpetuity of time; they never will make two unless an intelligence performs the act of adding them. Attentive intelligence alone has the power of carrying out the connections, and as soon as that attention relaxes, the connections dissolve. . . . The necessary connections which constitute the very reality of the world have no reality in themselves except as the object of intellectual attention in action.[65]

Weil's continuing analysis of necessity will show that not only does necessity constitute the mediation between the divine Creator and matter, but it also serves as the mediator between the physical and intellectual aspects of the human being. "Mathematical necessity is an intermediary between the whole natural part of man which is corporeal and psychical matter and the infinitely small portion of himself which does not belong to this world."[66] As we saw in the previous section, human beings possess something divine, the capacity to turn their attention toward what is other than this world. In the present context, we shall see that this capacity, exercised as love and consent, will be the key to understanding our relationship to the world around us.

When we contemplate and consider the connections and relationships that constitute necessity, Weil argues, we participate in an activity similar to the divine act of creation. "This virtue of intellectual attention makes it an image of the Wisdom of God. God creates by the act of thinking. We, by intellectual attention, do not indeed create, we produce no object, yet in our sphere we do in a certain way give birth to reality."[67] In a strong sense, then, the mathematics and science which arise from the proper investigation of necessity reflect both an activity and an obligation that participates in divinity. "This sensible universe in which we find ourselves has no other reality than that of necessity; and necessity is a combination of relations which fade away as soon as they are not sustained by a pure and lofty concentration on the part of the mind."[68] Undoubtedly, "the fact that the relationship which makes up the tissue of necessity is dependent upon the act of our attention makes it a thing belonging to us which we can love";[69] indeed, we shall see that only a science motivated by love for the material world can truly be a science for Weil. Yet there is much more to the story of necessity. Although necessity is the divinely imposed mediation between God and matter, a mediation that human knowers have a special capacity to know, the rigid necessity of the world often "seems an absolute and brutal master."[70] It is to this blind, mechanical aspect of necessity, the "mathematical progression of causes and effects which, from time to time, make of us a sort of formless jelly,"[71] that we now turn.

Affliction and Divine Providence

The fact that the natural world is governed by necessity, by an ensemble of mathematically precise laws and relationships, is so remarkable that it can (and should) lead one toward an admiration and love for the divine Creator who imposed this order on the world. "That this mathematical necessity should be the substance of the world—that is the sign of our Father, the witness that necessity was conquered from the beginning by a wise persuasion."[72] The very nature of necessity, however, makes it something that, more often than not, presents itself as something to resist and fear rather than something to love. For, "in its physical part, the soul is aware of necessity only as constraint and is aware of constraint only as pain."[73] The events of the natural world frequently reduce human beings to little more than things.

> A blind mechanism, heedless of degrees of spiritual perfection, continually buffets men hither and thither and flings some of them at the very foot of the Cross. . . . If the mechanism were not blind there would not be any affliction. Affliction is above all anonymous; it deprives victims of their personality and turns them into things.[74]

Weil is insistent that, above all, necessity is blind. There is no overall purpose to the laws and relationships that limit and structure the material universe—"things have causes and not ends."[75] Because we human beings are almost exclusively subject to necessity, just as are all other things in the world, we find ourselves at the mercy of the harshness and inexorability of natural phenomena.

It is this "metallic cold[ness] and hardness"[76] that we seek to overcome and manipulate to our purposes through science and technology oriented toward power. Weil points out, however, that when a human being is crushed by the blind contingency of natural events, we receive evidence that our attempts to harness necessity and turn it to our purposes are fundamentally misconceived. "So long as the play of circumstance around us leaves our being almost intact, or only half impaired, we more or less believe that the world is created and controlled by ourselves. It is affliction that reveals, suddenly and to our very great surprise, that we are totally mistaken."[77] There will be much of instructive value in the affliction to which humanity is frequently reduced by necessity. There is no mistaking the fact, however, that the brutal and hard face of necessity presents us with difficulties and obstacles that exceed the physical barriers presented by the contingency of events. There is a moral problem as well, particularly with the pain and suffering that necessity frequently imposes on the innocent. The key to understanding necessity and the problem of the suffering of the innocent is to situate necessity in the larger framework of transcendence that has been the focus of much of our study to this point.

> The effects of misfortune upon innocent souls are really unintelligible unless we remember that we have been created as brothers of the crucified Christ. The absolute domination throughout the whole universe of a mechanical, mathematical, absolutely deaf and blind necessity is unintelligible, unless one believes that the whole universe, in the totality of space and of time, has been created as the Cross of Christ.[78]

Constructing this perspective on the blind necessity of the world begins by addressing the apparent amorality of necessity, the fact that the effects of necessity are indiscriminately spread across the universe and all human beings, regardless of the moral or spiritual status of the persons impacted. What are we to make of "that blind impartiality characteristic of inert matter . . . that relentless regularity characterizing the order of the world, completely indifferent to men's individual quality, and because of this [the world is] so frequently accused of injustice"?[79] As Weil correctly notes, it is this aspect of the world that most frequently causes human beings to question the nature and intentions of the creator of the world. "The bitterest reproach that men make of this necessity is its absolute indifference to moral values. Righteous men and criminals receive an equal share of the benefits of the sun and of the rain; the righteous and the criminals equally suffer sunstroke, and drowning in floods."[80]

Perhaps the best example of this complaint, an example that Weil frequently raises when discussing the indifference of necessity, is the story of Job. In this familiar story Job, characterized by all (including God) as a righteous man, suffers a seemingly endless succession of disasters, all caused by natural events. The natural question, of course, is "Why?" The response of Job's wife and friends is that he must have committed some secret sin or sins to incur God's wrath; the underlying assumption behind this response is that God would obviously protect a truly righteous person from the disasters that have befallen Job. Job, however, steadfastly denies that any such secret transgressions exist. As we shall see, Weil considers the revelation that Job finally receives to be paradigmatic of what we must learn concerning the nature of the world around us and the necessity that seems so fundamentally oblivious to what we consider to be most important, our individual merit.

Weil argues that there is an entirely mistaken conception of divine providence and care at the heart of the response of Job's "comforters" to his distress and affliction, as well as at the heart of all who respond to the impartiality of necessity with questions as to why it rains on both the just and the unjust. This mistaken response is rooted in what Weil calls "the ridiculous conception of Providence as being a personal and particular intervention on the part of God for certain particular ends."[81] This conception presumes that while the necessity revealed in the order of the world is indeed reflective of God's general will, those who are righteous or specially favored will in some way be miraculously protected from the blind forces of necessity that frequently afflict others. The absence of such protection is then evidence of the lack of special favor. Weil speculates in *The Need for Roots* that this notion of a personal providence arose from the unfortunate adoption by the Roman Empire of Christianity as the state religion; God became a counterpart of the emperor, bestowing special favors on those deemed deserving.

The historical accuracy of Weil's speculations here is not important to our present purposes; what is of crucial importance is her insistence that the correct conception of divine providence, one that has roots in the Greek conception of the *kosmos*, is the conception of "an eternal and universal dispensation constituting the foundation of an invariable order in the world."[82] According to Weil, the belief in a personal providence is not only "incompatible with true faith . . . it is [also] incompatible with the scientific conception of the world."[83] By properly understanding that the inexorable order of the world is itself reflective of divine providence, we can begin to bring the scientific and theological conceptions of world order into line with each other. The impartiality of the natural order is not something to be overcome or something that we can expect to earn divine protection from; it is, in a real and profound sense, divine providence itself. "Divine Providence is not a disturbing influence, an anomaly in the ordering of the world: it is itself the order of the world; or rather it is the regulating principle of this universe. It is eternal Wisdom, unique, spread across the whole universe in a sovereign network of relations."[84] The necessary order of the world, expressed through mathematical regularity, is itself the will of God. "The sum of the particular intentions of God is the universe itself."[85]

If so, our task is to learn to read the order of the world as something to love, to embrace, rather than to resist or fear, even when this necessity afflicts us in ways that violate our expectations. This, Weil argues, is what Job's reaction toward his own affliction models.

> So it was that to Job, when once the veil of flesh had been rent by affliction, the world's stark beauty was revealed. The beauty of the world appears when we recognize that the substance of the universe is necessity and that the substance of necessity is obedience to a perfectly wise Love.[86]

How obedience and beauty are linked to properly understanding the nature of divine providence will be discussed below. At this point, just as in the case of interpersonal relationships between human beings, the beginning step toward a transformed understanding of the order of the world is to resist our natural tendency to view reality with ourselves at the center.

> We live in a world of unreality and dreams. To give up our imaginary position as the center, to renounce it, not only intellectually but in the imaginative part of our soul, that means to awaken to what is real and eternal, to see the true light and hear the true silence. A transformation then takes place at the very roots of our sensibility, in our immediate reception of sense impressions and psychological impressions.[87]

At the heart of this transformation is consent. Just as the key to bridging the gap between human beings who are unequal is the consent of the stronger to respect the autonomy of the weaker, so the key to seeing God in the necessity of the world is the choice to consent and be obedient to what we might, on a natural basis, seek to resist and overcome.

Consent and Obedience

How do we learn to consent to the necessary order of the world, when this order frequently impacts us, both physically and psychologically, in painful and destructive ways? We begin, Weil argues, by learning to view the blind necessity of the world as an example of perfect obedience, an example that we are not only to accept but also are to use as a model for our own attitudes toward necessity.

> Brute force is not sovereign in this world. It is by nature blind and indeterminate. What is sovereign in this world is determinateness, limit. Eternal Wisdom imprisons this universe in a network, a web of determinations. The universe accepts passively. The brute force of matter, which appears to us sovereign, is nothing else in reality but perfect obedience.[88]

This seemingly odd notion, that the necessity of the universe is to be understood as obedience to God, is rooted in several ideas that we have encountered previously. All that we know of matter is that it is what is indeterminate (*apeiron*) until it is brought by God within the limits of necessity. As matter governed and ordered by necessity, the things of this world keep within their assigned limits with respect to other things; by remaining within these limits, in Weil's manner of speaking, things are "obeying necessity." This obedience is blind, of course, in that matter has no choice as to whether it will remain within the limits imposed by necessity.

Necessity is reflective of the love and will of the Creator, hence, it is to be loved. Indeed, Weil claims, the Greek insight that the mathematically describable limits and boundaries of the universe are reflective of love rather than force served to energize their scientific enquiries.

> The thought which really enraptured the ancients was this: what makes the blind forces of matter obedient is not another, stronger force; it is love. They believed that matter was obedient to eternal Wisdom by virtue of the love which causes it to consent to this obedience. . . . The Greeks were thrilled to find in science a startling confirmation of this, and that was the cause of their enthusiasm for it.[89]

Weil argues that "it is our function in this world to consent to the existence of the universe. God is not satisfied with finding his creation good; he wants it also to find itself good."[90] This implies that we must learn to love the order of the world regardless of how it happens to effect us; "the order of the world is to be loved because it is pure obedience to God. Whatever this universe accords us or inflicts on us, it does so exclusively out of obedience."[91] Recalling first that the world does not "choose" to act in the way that it does and, second, that the very existence of the world requires the ultimate continuing mediating sacrifice of God, we can, by learning to love the necessity of the world, recognize "the infinite sweetness of obedience. For us, this obedience of things in relation to God is what the transparency of a window pane is in relation to light. As soon as we feel this obedience with our whole being, we see God."[92]

This requires, of course, that we embrace not only the beneficial and beautiful aspects of the world produced by the limits of necessity, but also that we embrace the pain, suffering, and affliction that the natural order produces in human life.

> One must tenderly love the harshness of that necessity which is like a coin with two faces, the one turned towards us being domination, and the one turned towards God, obedience. We must embrace it closely even if it offers its roughest surface and the roughness cuts into us.[93]

Human beings are things subject to the inexorable laws of necessity just as are all other things. There is, however, one difference that makes human beings more than just things subject to necessity. Although we cannot escape from necessity, we do have the capacity to choose how we will respond to the order of the world. We can either futilely resist and seek to control it, or we can consent to it, a consent energized by love for the Creator of whom the order of the world is a reflection.

> Man can never escape from obedience to God. A creature cannot but obey. The only choice given to men, as intelligent and free creatures, is to desire obedience or not to desire it. If a man does not desire it, he obeys all the same, perpetually, in as much as he is a thing subject to mechanical necessity. If he does desire it, he is still subject to mechanical necessity, but a new necessity is added to it, a necessity constituted by the laws pertaining supernatural things.[94]

The key to consenting to necessity, to choosing to embrace the order of the world even when it afflicts us, is once again the removal of ourselves from the center of the universe.

> From the time we renounce thinking in the first person, by consent to necessity, we see it from outside, beneath us for we have passed to God's side. The side which it turned to us before, and still presents to almost the whole of our being, the natural part of ourselves, is brute domination. The side which it presents after this operation, to the fragment of our mind which has passed to the other side, is pure obedience. We have become sons of the home, and we love the docility of this slave, necessity, which at first we took for a master.[95]

We saw in our discussion of relationships between human beings earlier in this chapter that the human capacity to turn one's attention from the natural toward what is transcendent is a supernatural capacity, something we choose to use when we recognize the spark of divinity in each person, regardless of their contingent inequalities. It is this same supernatural capacity for consent that makes it possible for human beings to embrace the order of the world. Just as Christ consents to the suffering required for creation to exist, human beings participate in that suffering by consenting to the existence of the creation in all of its beauty as well as its pain.

> The faculty which does not belong to this world is the faculty of consent. Man is free to consent to necessity, or not. . . . No mover, no motive can be sufficient for such a consent. This consent is madness, man's own particular madness, the madness that belongs to man, like Creation, like the Incarnation, together with the Passion, constitute God's own madness. . . . The true response to the excess of divine love . . . consists only in consenting to the possibility of being destroyed, that is to say, in the possibility of total disaster, whether that disaster actually happens or not.[96]

Once again, the model for such an embracing of the world in all of its beauty and danger is Job, who claimed in the midst of affliction that "though He slay me, yet will I trust in Him."[97] Such consent is mad and absurd, because there are no guarantees—"We cannot see the fruits of this consent."[98] Yet when a human being does consent to necessity, that human being participates in the heart of God.

> The whole universe is a compact mass of obedience. This compact mass is sprinkled with points of light. Each one of these points is the supernatural part of the soul of a reasonable creature who loves God and who consents to obey. . . . The consent is supernatural love, it is the Spirit of God in us. . . . The consent to obey is mediator between blind obedience and God. The perfect consent is that of the Christ. Our consent can only be a reflection of that of the Christ. . . . We, in so far as it is granted us to imitate Christ, have this extraordinary privilege of being, to a certain degree, mediators between God and His own creation.[99]

Recalling that it is the human intellect, in its reproduction of reality under ordered mathematical principles, that forms the object (the order of the world) to which we may choose to consent, it is truly in the embracing of reality in all of its aspects that we express our love of God more completely.

> Being subject in this world to her [necessity's] tyranny, we have only to choose God for our treasure, and put our heart with it, and from that moment we shall see the other face of the tyranny, the face which is pure obedience. We are the slaves of necessity, but we are also the sons of her Master.[100]

Beauty

In the introduction to this study, we noted Subrahmanyan Chandrasekhar's observation that "beauty is that to which the human mind responds at its deepest and most profound."[101] As we approach the conclusion of our investigation of the relationship of human beings to the things of the natural order, we return with Simone Weil to the beauty of the world. She would have agreed with Chandrasekhar's observation that beauty serves as a powerful motivating influence in all human beings.

> A sense of beauty, although mutilated, distorted, and soiled, remains rooted in the heart of man as a powerful incentive. It is present in all the preoccupations of secular life. If it were made true and pure, it would sweep all secular life in a body to the feet of God; it would make the total incarnation of the faith possible.[102]

Consenting to the necessity that mediates between God and matter brings a human being to a revelation. As Job discovered, the very necessity that reduced him to a state of affliction is precisely the necessity that is responsible for the beauty of the world.

> Every sort of grief, and above all every sort of disaster well endured, makes the passage to the other side of that door possible, makes the true face of harmony seen. . . . That is what the end of the book of

Job reveals. Job, at the end of his distress, which despite appearances he has perfectly well endured, receives the revelation of the beauty of the world.[103]

When we encounter the beauty of the world with an understanding of both the divine love and suffering responsible for it as well as the human love and affliction required to respond to it fully, we are encountering God.

> God created the universe, and his Son, our first-born brother, created the beauty of it for us. The beauty of the world is Christ's tender smile for us coming through matter. He is really present in the universal beauty. The love of this beauty proceeds from God dwelling in our souls and goes out to God present in the universe. It also is like a sacrament.[104]

Weil suggests that we cannot, as has often been attempted, use the beauty of the world as a foundational element in an attempt to prove the existence of God. Faith in the existence of God comes first, but experiencing and properly understanding the beauty of the world provides us with a powerful confirmation of that faith. "God is not proved by the goodness of the universe, but the goodness of the universe by God; or rather, it is a matter of faith. But the universe is beautiful, even including evil, which, as part of the order of the world, has a sort of terrible beauty. We feel it."[105] Elsewhere, Weil writes that "we must have faith that the universe is beautiful on all levels,"[106] even when the necessity that produces the world's beauty afflicts us. This is why understanding the mediating activity responsible for the existence of the universe is crucial for engaging beauty properly. If we begin with beauty, we will fail to understand the equal necessity of the pain and affliction that seem, on the surface, to be contrary to beauty. If we understand, however, that the necessity of the world is in a real sense equivalent to the beauty of the world, we will be able to have a better and better grasp on how all features of reality are expressions of divine love. "Whenever we have some pain to endure, we can say to ourselves that it is the universe, the order and beauty of the world, and the obedience of creation to God which are entering our body."[107]

Undoubtedly, "beauty is a mystery; it is what is most mysterious here below."[108] Given our considerations to this point, however, what does the beauty of the world, produced by the imposition of mathematical, necessary limits on matter, reveal to us concerning the nature of God? Weil reminds us that we must consider necessity, as well as divine providence, as impersonal, an aspect of God's dealings with the world that Christianity has largely lost sight of. With this in mind, we see that beauty reveals first the impartial and impersonal aspects of God's nature, as Weil suggests in the following fragments from her Marseilles notebooks:

> The beauty of the world proves that there is a God who is at the same time personal and impersonal, and neither just the one nor the other. Author and order. Necessity also (mathematical and mechanical relationships) represents an order without an author. Mathematics as *metaxu* leading towards the impersonal aspect of God.[109]

Yet, at the same time, the beauty of the world is what is most likely to draw the individual human being toward God in a way that reveals the divine desire to contact each of us on a personal basis. "The soul's natural inclination to love beauty is the trap God most frequently uses in order to win it and open it to the breath from on high. . . . The longing to love the beauty of the world in a human being is essentially the longing for the Incarnation."[110] Indeed, Weil reminds us, we as perceiving and thinking creatures have a crucial role in the creation of beauty itself.

> The beauty of the world is not an attribute of matter in itself. It is a relationship of the world to our sensibility, the sensibility that depends upon the structure of our body and our soul. . . . We must have faith that the universe is beautiful on all levels, and more generally that it has a fullness of beauty in relation to the bodily and psychic structure of each of the thinking beings that actually do exist and of all those that are possible. It is this very agreement of an infinity of perfect beauties that gives a transcendent character to the beauty of the world. Nevertheless the part of this beauty we experience is designed and destined for our human sensibility.[111]

Ultimately, all investigations into truth and beauty are, due to our human imperfections and limitations, incomplete reflections of the unity and wholeness of goodness and beauty.

> All goods in this world, all beauties, all truths, are diverse and partial aspects of one unique good. Therefore they are goods which need to be ranged in order. Puzzle games are an image of this operation. Taken all together, viewed from the right point and rightly related, they make an architecture. Through this architecture the unique good, which cannot be grasped, becomes apprehensible.[112]

Parts of this architecture are the geometrical and mathematical truths that have been the frequent focus of this present study. In many ways, the investigation of how mathematical limits fashion matter into beauty brings the human person closest to God of all human investigations, for "there is here below but one single beauty, that is the Beauty of the World. All other beauties are reflections of that one."[113] The beauty of the world is so important to the possibility of human contact with God that Weil suggests that we ignore it or misrepresent it at our peril.

> The brightness of stars, the sound of sea waves, the silence of the hour before dawn—how often do they not offer themselves in vain to men's attention? To pay no attention to the world's beauty is, perhaps, so great a crime of ingratitude that it deserves the punishment of affliction.[114]

Much is at stake, for only a renewed love for the universe, a love such as that expressed through Greek art, science, and mathematics, will make possible a even larger transformation.

> Between God and these incomplete, unconscious, often criminal [ways of] searching for beauty, the only link is the beauty of the world. Christianity will not be incarnated so long as there is not joined to it the Stoic's idea of filial piety for the city of the world, for the country of here below which is the universe.[115]

"A Preparation for Deliverance"

The object of science is the presence of Wisdom in the universe, Wisdom of which we are the brothers, the presence of Christ, expressed through matter which constitutes the world.

—Simone Weil, "Forms of the Implicit Love of God," *WG*, 169

In the introduction to this study, we noted Simone Weil's identification of "a task worth accomplishing," that task being "to restore to science as a whole, for mathematics as well as for psychology and sociology, the sense of its origin and veritable destiny as a bridge leading toward God."[1] The overriding purpose of our considerations has been to identify with as much precision as possible the various features of Weil's thought that come to bear on this project; we have seen that although she never did more than sketch the outline of such a restored science, the foundational elements that must be in place for such a project to even make sense are carefully considered throughout her writings.

Of all the various definitions of science that Weil provides us in her writings, the above suggestion that science is what investigates "the presence of Wisdom in the universe" is particularly useful for our purposes, since this presence unifies the seemingly disparate elements that constitute the beauty of the world. Weil describes this unification in the following passage a few pages from the conclusion of *The Need for Roots.* "The order of the world is the same as the beauty of the world. . . . It is one and the same thing, which with respect to God is eternal Wisdom; with respect to the universe, perfect obedience; with respect to our love, beauty; with respect to our intelligence, balance of necessary relations; with respect to our flesh, brute force."[2] The same necessity that reveals the obedience of matter, bears on our physical selves frequently by producing pain and suffering, and is "resplendent in beauty"[3] when engaged by love, is revealed by the intelligence in scientific enquiry as "the balance of necessary relations." All of these are the various faces of "eternal Wisdom" as it effects creation.

The development of a science rooted in love rather than power is, of course, more than a matter of taste or personal preference for Weil. We have seen that she believed that "the modern conception of science is re-sponsible . . . for the monstrous conditions under which we live, and will, in its turn have to be transformed, before we can hope to see the dawn of a better civilization."[4] This transformation, which will require "another effort comparable to that of Eudoxus,"[5] cannot occur "unless the Christian tradi-tion comes alive in us again."[6] In what ways might a transformed science, rooted in a renewed transcendent framework, serve to bridge the contempo-rary gap between religion and science as well as to illuminate the connection between love and the search for truth?

EQUILIBRIUM

In an unfinished and undated essay entitled "Foundation of a New Sci-ence," Weil provides us with one of her few attempts to outline what a sci-ence built on the foundations of love rather than power might actually look like. She begins by drawing our attention yet again to the crucial importance of limitation, proportion, and restraint in our understanding of the universe.

> Limitation is the law of the manifested world. . . . Man, who is of the world and who has a part in God, puts the unlimited and absolute into the world, where they are error; this error is suffering and sin, and all human beings, even the most ignorant, are torn by this con-tradiction. . . . Deliverance consists in reading limit and relation in

all sensible phenomena without exception, with the same clarity and immediacy as a meaning in a printed text. The significance of a true science is to constitute a preparation for deliverance.[7]

The dream (and delusion) of science over the past several centuries has been of unlimited progress, a continuing and increasing mastery of the world around us through increased knowledge and technology. Weil, whose life was bracketed by the First and Second World Wars, had firsthand evidence of the dangers of a science dedicated to nothing but power and technology; the dangers she warned of have become more and more of a reality in the decades since her death. Such a science contains little or no awareness of limits and seeks only to expand domination as far as possible. Such hopes are futile, Weil argues, on a practical basis because

> no matter how much use is made of algebra and instruments, science will always depend largely upon man's intelligence and physique, which are limited and do not become less so with the passing of centuries; so it is absurd to believe that science is capable of unlimited progress. It is limited, like all human things, except that point in man which is assimilated to God.[8]

More important, however, is to eliminate the illusion of limitlessness, since "in man, injustice consists in disregarding limits."[9] As we have seen, one of the central themes of Weil's meditations on science and mathematics is that the order of the world is produced by the imposition of a network of limits, of necessity, by the Creator on matter. Truly "limitation is the law of the manifested world." Hence, a science that sees in such limits something to contemplate and to love rather than restraints to overcome through increased power will look quite different than our present science. Such a science will serve as a "preparation for deliverance" from our attempts to manipulate or master the world.

Weil's simple suggestion in "Foundation of a New Science" is that science must be viewed as a means to something else, specifically loving God, rather than an end in itself. In other words, "what is needed now is not to try to extend it [science] further, but to bring thought to bear upon it."[10] This contemplation begins with the search in our scientific investigations, not for ways to increase domination, but rather for instances of limitation in action, instances of equilibrium.

> Equilibrium, in so far as equilibrium defines limits, is the essential idea of science; by means of this idea every change, and therefore every phenomenon, is considered as a rupture of equilibrium, linked with all other changes through the compensation of successive ruptures of equilibrium; and this compensation makes all disequilibria an image of equilibrium, all changes an image of the motionless, and time an image of eternity.[11]

Weil is clear that the equilibrium of the universe, expressive of the limits that make the universe possible, is something that we must believe in at the outset; it is not something that we can discover in its entirety through experimentation.

> It would be necessary, in order to perceive an image of equilibrium in the indefinite succession of ruptures of equilibrium, to embrace the totality of the universe and of time; and this is not granted to man, whose thought, in so far as it relates to objects, is limited. Man would be incapable of forming a concrete conception of equilibrium, and therefore of limit, and therefore of the unlimited, and he would be incapable in a general way of thought, unless he were given images of equilibrium on his own scale. This is not a necessity, however, but a grace; a grace indistinguishable from the grace by which man exists.[12]

Our imperfections, finitude, and limitations make it impossible for us to confirm experimentally the hypothesis that the totality of the universe is expressive of limit, balance, and equilibrium. This is something that must be taken on faith. We could not even make sense of the notion of limit at all, however, unless we possessed the capacity to recognize it on a limited scale. The fact that we can recognize it is yet another sign of divine grace.

Hence, when we do encounter balance and proportion, whether in pure mathematics or in the tracing of the mathematical limits of the natural world, we have encountered something reflective of God. "We have understood something when we have defined an equilibrium, and limits in relation to this equilibrium, and relations of compensation linking successive ruptures of equilibrium. This is also true for studies of social life and of the human soul, and only in this way can they be sciences."[13] We saw in the previous chapter that, according to Weil, "we must have faith that the universe is beautiful on all levels,"[14] even when the necessity that produces the world's beauty afflicts us. Similarly, we must begin our investigations of the world with the conviction that limits and equilibria exist on all levels of reality, even though we are frequently incapable of identifying them fully. This conviction can serve as a corrective to our natural tendency to deny or ignore limitations, believing that we are "the measure of all things."

Such a conviction is, of course, contrary to what contemporary science takes to be the essence of scientific enquiry. Contemporary science tells us that value and fact do not belong together; to assume the harmony and equilibrium of the universe before it is derived from empirical data is to illicitly impose value on fact. Weil argues, however, that no investigation is value free. All we can do is choose what values we will adopt in the construction of the metaphysical foundations and framework of our science; the burden of this study has been that rooting our science in values such as beauty and love is a meaningful alternative to building it on values such as power and dominance.

SCIENCE AND RELIGION

In many of the texts that we have studied in the previous pages, Simone Weil has sought to establish that for the ancient Greeks, science was something closely connected to human endeavors that, since the Scientific Revolution, have been split from science almost entirely. The belief that religious belief and scientific rigor are incompatible, for instance, has become so common that, among both religious believers and dedicated believers in the power of science, the belief has attained the status of a truism. This disconnect leads either to the removal of religion from the realm of legitimate human intellectual endeavor or to the notion that religion and science are not really in conflict because they are dealing with entirely different subject matters. Just as the rules of baseball and basketball are not in conflict with each other because they are intended to apply to entirely different games, so religion and science are different "games" played by different rules of evidence and procedure.

Weil's analysis of the disconnect between science and religion is intended not only as a diagnosis of why these activities came to be considered as incompatible but also as a sketch of how religion and science can be unified once again. She begins by suggesting that the natural state of affairs is for science and religion to be unified in both energy and focus. "The savant's true aim is the union of his own mind with the mysterious wisdom eternally inscribed in the universe. That being so, how should there be any opposition or even separation between the spirit of science and that of religion? Scientific investigation is simply a form of religious contemplation."[15] If this was the case in ancient Greece, it is certainly not the case any longer. As Weil writes, for contemporary Christians, "the absolute incompatibility between the spirit of religion and that of science, to both of which they cling, leaves the soul in a permanent state of secret, unacknowledged uneasiness."[16]

The contemporary view of divine providence, of a special favor bestowed on those whom God favors, has produced an almost schizophrenic situation in which the believer must compartmentalize religion and science in order to avoid facing their entire incompatibility.

> Christians who, under the influence of education and surroundings, carry within them this conception of Providence, also carry within them the scientific conception of the world, and that divides their minds into two water-tight compartments: one for the scientific conception of the world, the other for the conception of the world as being a field in which God's personal Providence is exercised. This makes it impossible for them really to think either the one or the other. The second one, moreover, will not bear serious scrutiny. Unbelievers, not being inhibited by any motives of reverence, detect easily enough the ridiculous aspect of this personal and particular form of Providence, and religious faith itself is, on account of it, made to seem ridiculous in their eyes.[17]

For many, "this is why religion is nowadays something we relegate to Sunday morning. The rest of the week is dominated by the spirit of science."[18] We end up with a religion that is rendered ridiculous by its view of God as a capricious Being who favors the few by violating the laws of nature and a science that is severed from its true end, the love of God.

If science and religious contemplation are to become equivalent, we must return to a strong sense of the impersonal aspects of God. In Weil's estimation, divine providence and the order of the world are the same and are the proper object of scientific inquiry. It is not enough, however, simply to assent intellectually to the proposition that God's perfect will is the order of the world. We must embrace that order through love; only then will science and religion be unified through the spirit of truth. "The spirit of truth is nowadays almost absent from religion and from science and from the whole of thought. . . . The remedy is to bring back again among us the spirit of truth, and to start with in religion and science; which implies that the two of them should become reconciled."[19]

One of Simone Weil's most direct criticisms of contemporary science is that "present-day scientists have nothing in their minds, however vague, remote, arbitrary, or improbable, which they can turn towards and call it truth."[20] This statement would amaze most scientists, since what is truth, if not the proper interpretation of empirically gathered and tested data? If the world is nothing but a mechanism operating according to rigid and blind laws, then perhaps accurate description is all we can expect. In Weil's framework, however, the mechanism of necessity becomes something to love; the poverty of contemporary science is that because of its rejection of any transcendent end toward which it is but a means, there is in this science "nothing . . . that a human mind can love."[21]

Clearly, the acquisition of facts and knowledge is an integral part of science. The mere accumulation of knowledge, however, never amounts to truth, as Weil argues in the following passage from *The Need for Roots*:

> The acquisition of knowledge causes us to approach truth when it is a question of knowledge about something we love, and not in any other case. Love of truth is not a correct form of expression. Truth is not an object of love. . . . A truth is always the truth with reference to something. Truth is the radiant manifestation of reality. . . . To desire truth is to desire contact with a piece of reality. To desire contact with a piece of reality is to love.[22]

Why must science be built on a foundation of love for the order of the world and all that order implies for Weil? Because one can only pursue the truth about something that one loves. Greek science and mathematics are expressions of and attempts to understand the beauty and order of the world, an understanding that can only be motivated by love. When this love is absent, the truth is absent, and the vacuum created by the absence of truth is filled by power and manipulation. "Today, science, history, politics, the organization of labor, religion even . . . offer nothing to men's minds except brute force. Such is our civilization. It is a tree which bears the fruit it deserves."[23] Corrupt trees bear only rotten fruit.

This is why Weil's vision of a transformed culture, outlined in *The Need for Roots,* necessarily includes a transformed science that can be reunited with religious and artistic energies. Above all, we must be drawn toward the truth in whatever endeavor by love alone.

> Pure and genuine love always desires above all to dwell wholly in the truth whatever it may be, unconditionally. Every other sort of love desires before anything else means of satisfaction, and for this reason is a source of error and falsehood. Pure and genuine love is in itself spirit of truth. It is the Holy Spirit. The Greek word which is translated spirit means literally fiery breath, breath mingled with fire, and it represented, in antiquity, the notion which science represents today by the word energy. What we translate by "spirit of truth" signifies the energy of truth, truth as an active force. Pure love is this active force, the love which will not at any price, under any condition, have anything to do with either falsehood or error.[24]

When a human being is committed to the truth through love at all costs, refusing to be comforted by illusion or misguided desire, the promise of resurrection becomes a reality.

> They alone will see God who prefer to recognize the truth and die, instead of living a long and happy existence in a state of illusion. One must want to go towards reality; then, when one thinks one has found a corpse, one meets an angel who says: "He is risen."[25]

Notes

INTRODUCTION. *Kosmos* and Crisis

1. Czeslaw Milosz, *Emperor of the Earth: Modes of Eccentric Vision* (Berkeley: University of California Press, 1977), 89.

2. The most extended treatment of this aspect of Weil's thought is contained in Eric O. Springsted, *Christus Mediator: Platonic Mediation in the Thought of Simone Weil* (Chico, CA: Scholars Press, 1983). I will make occasional reference throughout this present study to *Christus Mediator*, drawing attention both to similarities and differences in our treatment of parallel topics.

3. In *FW*, 31–88. This important, although largely neglected, early work will be considered in some detail in chapter 1.

4. *NR*, 179–288.

5. Ibid., vii.

6. Richard Dawkins, *Unweaving the Rainbow* (Boston: Mariner Books, 1998), 63.

7. "The Pythagorean Doctrine," *IC*, 172.

8. *FLN*, 71.

9. "Science and Perception in Descartes," *FW*, 46.

10. "Divine Love in Creation," *IC*, 103.

11. *FLN*, 29.

12. Ibid., 79.

13. "The Romanesque Renaissance," *SE*, 52.

14. "Fragment: Foundation of a New Science," *SN*, 80.

15. "Forms of the Implicit Love of God," *WG*, 169.

16. "Divine Love in Creation," *IC*, 103.

17. *NB*, 248.

18. Ibid., 440.

19. "Human Personality," *SE*, 14.

20. Ibid., 34.

21. *FLN*, 80.

22. Weil frequently contrasts obedience with seeking to gain power and control as she seeks to contrast her vision of science with the vision that arose from the Scientific Revolution. It is worth remembering, however, that even Francis Bacon said that "man has command over nature only by obeying it," as Weil reminds us at *LP*, 89.

23. *NR*, 227.
24. Ibid., 229.
25. "Classical Science and After," *SN*, 30.
26. "The Power of Words," *SE*, 156.
27. "Classical Science and After," *SN*, 30.
28. "Reflections on Quantum Theory," *SN*, 63–64.
29. *NR*, 229.
30. Ibid.
31. Ibid., 229–30.
32. *SL*, 90.
33. *NR*, 227.
34. Ibid., 226–27.
35. Ibid., 232–33.
36. *SL*, 131.
37. "Reflections on Quantum Theory," *SN*, 64.

CHAPTER ONE.   Science and Work

1. "Classical Science and After," *SN*, 3.
2. *FW*, 32. All references to *FW* in this chapter are to "Science and Perception in Descartes."
3. Ibid., 31.
4. *LP*, 72.
5. Ibid., 32.
6. As we shall see, Weil combines this commitment to self-sufficiency and suspicion of collective authority with a deep appreciation for the importance of social context and collective effort.
7. *FW*, 33.
8. Ibid. Weil uses the phrase "modern science" here in reference to science since Descartes. In her later writing on science, she uses "modern science" interchangeably with "classical science," both referring roughly to science from the Scientific Revolution to the beginning of the twentieth century. She refers in her later writings to science after 1900 as "contemporary science."
9. "Human Personality," *SE*, 13.
10. *FW*, 33.

11. Ibid.

12. Ibid., 33–34.

13. *LP*, 71.

14. Weil's later published writings and notebooks are filled with deeper and deeper concerns about the nature of algebra and how contemporary scientists use it. This theme will be picked up in chapter 2.

15. *FLN*, 31.

16. *FW*, 34.

17. *LP*, 85.

18. Ibid., 85–86.

19. Ibid., 86.

20. "Wave Mechanics," *SN*, 76.

21. *FW*, 35.

22. Ibid., 43.

23. Ibid., 36–37.

24. Ibid., 38; AT 9:93. All quotations from Descartes are translated by Dorothy Tuck McFarland and Wilhelmina Van Ness in *FW*. The second citation is to volume and page in the standard edition of Descartes' works, *Oeuvres de Descartes,* ed. Charles Adam and Paul Tannery, 12 vols. (Paris: Vrin, 1897–1910).

25. Richard D. McKirahan, Jr., *Philosophy Before Socrates* (Indianapolis: Hackett Publishing Company, 1994), 26.

26. *FW*, 39.

27. Ibid., 40.

28. Ibid., 43.

29. Ibid., 44; AT 6:61.

30. Ibid., 44.

31. Ibid., 47; AT 10:374–75.

32. Ibid., 47; AT 10:360.

33. Ibid., 53.

34. Ibid., 48.

35. Ibid., 49.

36. Ibid., 50.

37. Ibid.; AT 3:692.

38. Ibid.

39. Ibid., 51. Under the traditional reading of Descartes, the imagination is a hindrance to any attempt to find certainty in knowledge; under Weil's reading of Descartes, the imagination is indispensable to human knowledge although, as we shall see, its use can be abused and can exceed its proper boundaries.

40. Ibid., 54.

41. Ibid.

42. Ibid., 54–55.

43. Ibid., 55–56.

44. Weil's approach to this question is somewhat different in "Science and Perception in Descartes" than in *Lectures on Philosophy,* notes from classes given four years later. Peter Winch has effectively argued that the approach in the latter text is significantly different and more mature than that in her dissertation (*LP,* 1–23). For the purposes of continuity I will focus primarily on her approach in "Science and Perception in Descartes," but will draw attention on occasion to her approach in *LP* when the latter approach indicates a significant change in her thinking.

45. *FW,* 56.

46. Ibid., 57.

47. Ibid., 58.

48. Ibid.

49. Ibid., 58–59.

50. Ibid., 59.

51. *LP,* 5.

52. *FW,* 62.

53. Ibid., 63.

54. Ibid., 64.

55. Ibid., 69.

56. Ibid.

57. Ibid., 70.

58. *LP,* 52.

59. *FW,* 72.

60. Ibid., 73.

61. Ibid., 74.

62. *LP,* 73.

63. Ibid., 71.

64. *FW,* 75.

65. Ibid.

66. *LP,* 31–32.

67. *FW,* 77–78.

68. Ibid., 78.

69. Ibid.

70. Ibid.

71. Ibid., 79.

72. *LP,* 80.

73. *FW,* 82.

74. Ibid., 80.

75. "Classical Science and After," *SN,* 41.

76. *FW,* 81.

77. Ibid., 84.

78. Quoted by Weil in *FLN,* 25.

79. *FW*, 79.
80. Ibid., 81.
81. *NR*, 288.
82. *FW*, 84.
83. Ibid.
84. Ibid., 85–86.
85. Ibid, 85.
86. *LP*, 119.
87. *SL*, 20.
88. *FW*, 81.
89. *FLN*, 44.

CHAPTER TWO.   Classical and Contemporary Science

1. "Reflections on Quantum Theory," *SN*, 62.
2. Ibid.
3. Ibid., 63–64.
4. *NR*, 243.
5. *SN*, 3–48.
6. "Classical Science and After," *SN*, 6.
7. Ibid., 8.
8. "Reflections on Quantum Theory," *SN*, 52.
9. "Classical Science and After," *SN*, 5–6.
10. Ibid., 8.
11. Ibid.
12. Ibid., 8–9.
13. Ibid., 9.
14. Ibid., 10.
15. Iris Murdoch, *The Sovereignty of Good* (London: Routledge, 1970), 103.
16. "Scientism: A Review," *SN*, 68.
17. Ibid.
18. "Variant: Meditation on Greek Science," *SN*, 44.
19. Ibid.
20. "Classical Science and After," *SN*, 6.
21. Ibid.
22. Ibid., 7.
23. Ibid., 11.
24. Ibid., 11–12.
25. Ibid., 16–17.
26. *NR*, 233.
27. "Classical Science and After," *SN*, 15–16.

28. *NB,* 34.

29. "Reflections on Quantum Theory," *SN,* 59–60.

30. "Classical Science and After," *SN,* 10.

31. Ibid., 23.

32. Ibid., 3.

33. Ibid. This offhanded remark can be justified in Descartes' case by considering sections 28–32 of part 2 of his *Principles of Philosophy.* In the case of Newton, Weil is apparently inferring from the celebrated scholium to the definitions of book 1 of the *Principia* that Newton would not have worked so hard to distinguish absolute from relative time, space, and motion if he thought that the notion of motion and rest being entirely relative to a given system of reference is "a patent absurdity."

34. "Reflections on Quantum Theory," *SN,* 49.

35. The following pages are by no means intended as a full introduction to quantum theory; rather, my intent is simply to highlight several important aspects of quantum mechanics that were of particular interest and concern to Weil at their stage of development in the early 1940s. There are several good introductions to quantum theory for the educated reader in print, with more appearing all the time.

36. "Classical Science and After," *SN,* 5.

37. "Reflections on Quantum Theory," *SN,* 50.

38. "Classical Science and After," *SN,* 3.

39. "Reflections on Quantum Theory," *SN,* 49.

40. *SL,* 89.

41. "Scientism: A Review," *SN,* 70.

42. Ibid.

43. "Reflections on Quantum Theory," *SN,* 60.

44. *NB,* 75.

45. "Classical Science and After," *SN,* 22.

46. Ibid.

47. Ibid., 26.

48. Ibid.

49. "Reflections on Quantum Theory," *SN,* 50.

50. "Classical Science and After," *SN,* 22–23.

51. Ibid., 5.

52. Ibid., 27.

53. "Reflections on Quantum Theory," *SN,* 53.

54. "Classical Science and After," *SN,* 23.

55. Ibid., 22.

56. *SL,* 3.

57. "Reflections on Quantum Theory," *SN,* 54.

58. Ibid.

59. Ibid.

60. Ibid., 49–50.

61. Ibid., 54.

62. *LP*, 117–18.

63. "Classical Science and After," *SN*, 23.

64. Gary Zukav, *The Dancing Wu Li Masters* (New York: Bantam Books, 1980), 208.

65. *FLN*, 30.

66. "Reflections on Quantum Theory," *SN*, 54–55.

67. *NB*, 79.

68. "Wave Mechanics," *SN*, 75.

69. "Reflections on Quantum Theory," *SN*, 55.

70. Ibid.

71. Ibid., 63.

72. Max Planck, *Initiations à la physique* (Paris: Flammarion, 1941), quoted in "Reflections on Quantum Theory," *SN*, 56.

73. *NR*, 247.

74. "Reflections on Quantum Theory," *SN*, 58.

75. Ibid., 58–59.

76. Ibid., 57–58.

77. Ibid., 63.

78. Ibid., 63–64.

79. Ibid., 64.

80. Ibid.

81. *NB*, 595.

CHAPTER THREE.   Monochords and Bridges

1. "Science and Perception in Descartes," *FW*, 32.

2. Ibid., 33.

3. *NR*, 233.

4. "Variant: Meditation on Greek Science," *SN*, 47.

5. *SL*, 118.

6. "Classical Science and After," *SN*, 16.

7. There are a number of books available that provide a useful orientation to the Pythagoreans and Pythagoreanism. I have found Richard D. McKirahan, Jr., *Philosophy Before Socrates* (Indianapolis: Hackett Publishing Co., 1994), and Kenneth Sylvan Guthrie, *The Pythagorean Sourcebook and Library,* introduced and edited by David R. Fideler (Grand Rapids: Phanes Press, 1987), to be particularly helpful.

8. "Classical Science and After," *SN*, 21.

9. "The Pythagorean Doctrine," *IC*, 153.

10. Ibid., 154.

11. Ibid., 160.

12. Ibid., 155.

13. *NB,* 392.

14. *NR,* 279. This passage, from which the title of our present study is taken, raises a puzzle in that it is not always clear in Weil whether the mathematics which is "the stuff of which the world is woven" is "the threads" from which that world is woven, or the resulting cloth that is woven. I believe, as we shall see in the discussion below, that Weil would consider mathematics to be both the weave of the world and the stuff of which the world is woven.

15. "The Pythagorean Doctrine," *IC,* 153.

16. *FLN,* 88.

17. *NB,* 191.

18. "The Pythagorean Doctrine, *IC,* 153.

19. Ibid., 159.

20. Quoted in Guthrie, 21.

21. *FLN,* 250.

22. For a further discussion of the limited and unlimited, see pp. 63–67 of Diogenes Allen and Eric O. Springsted, "The Concept of Reading and the Book of Nature," in *Spirit, Nature, and Community* (Albany: State University of New York Press, 1994). Many of the essays in *Spirit, Nature, and Community,* coauthored by Allen and Springsted, are directly relevant to our present study.

23. Aristotle, *Metaphysics* I.5, 986a23.

24. *FLN,* 16.

25. Ibid., 126.

26. *NR,* 274.

27. Plato, *Epinomis* 990d, quoted in "The Pythagorean Doctrine," *IC,* 156–57.

28. "The Pythagorean Doctrine," *IC,* 162.

29. The concept of mediation is crucial in Weil's thought, extending well beyond the scope of the geometrical and mathematical context of our present study. Springsted's *Christus Mediator* is an extended study of this aspect of Weil's thought; in Springsted's own words, "mediation in the writings of Simone Weil is essentially a means of dealing with incommensurate and contrary realities in such a way that, without destroying their individual integrity, they can be harmoniously united" (8). Although we will frequently encounter Weil's use of mediation in our present study, *Christus Mediator* provides an exhaustive treatment of the topic.

30. "The Romanesque Renaissance," *SE,* 51.

31. François Lassere, *The Birth of Mathematics in the Age of Plato* (Cleveland: Meridian Books, 1966), 83–84.

32. McKirahan, 115.

33. Ibid.

34. Rush Rhees, *Discussions of Simone Weil,* edited by D. Z. Phillips (Albany: SUNY Press, 2000).

35. The important and complex issues arising from Rhees' Wittgensteinian/ analytic reading of Weil can only be briefly introduced in this present study. *Discussions of Simone Weil* consists primarily of unpublished letters and notes, primarily written in the late 1960s, that were collected and published after Rhees' death in 1989. Some of Rhees' excerpted letters were written to Peter Winch, whose *Simone Weil: The Just Balance* (Cambridge: Cambridge University Press, 1989) picks up and advances a number of the philosophical themes that arise in Rhees' reflections. Allen and Springsted respond to and develop a number of Winch's points in *Spirit, Nature, and Community*, particularly in "The Concept of Reading and the Book of Nature" (53–76) and "Winch on Weil's Supernaturalism" (77–93). See also Eric O. Springsted, "Spiritual Apprenticeship," in *Cahiers Simone Weil* 25 (2002), 325–44.

36. Ibid., 64.

37. Ibid., 92.

38. Ibid., 85.

39. Rhees' concern is a reflection of his commitment to "ordinary language" philosophy that first and foremost focuses its philosophical attention on a critical analysis of the use of language rather than a priori metaphysical systems. Sharing this same commitment, Winch points out that we must remember that "the fact that one and the same phrase is used in two different contexts is poor evidence that the same sort of discussion is going on." Otherwise, "the kinds of consideration she [Weil] appeals to must appear quite astounding or, not to put too fine a point on it, crazy" (192).

40. Ibid., 90.

41. Ibid., 56.

42. Ibid., 65.

43. Ibid., 88.

44. Ibid., 67.

45. "I feel like complaining that she *mixes up* philosophy and religious meditation, and writes as if she were not even aware that she was doing so." Ibid., 86.

46. Ibid.

47. Ibid., 64.

48. Ibid., 86.

49. "Variant: Meditation on Greek Science," *SN*, 47.

50. "God in Plato," *SN*, 90. The importance of what Weil means by "bridges" will be discussed in the next section.

51. "The Pythagorean Doctrine," *IC*, 191.

52. Ibid., 171.

53. Rhees would perhaps argue that the issue here is not one of competing metaphysical systems, but rather his trying to avoid having to read Weil as being a metaphysician at all. One way of interpreting Weil would be as in the tradition of philosophical metaphysicians who develop a priori frameworks of reality and then

insert the phenomena of human experience and reality into this framework. Rhees is committed to developing understanding from the phenomena (particularly of language and concept formation first), without prior metaphysical frameworks and commitments. This analytic suspicion of a priori metaphysics is important, but is itself reflective of a priori assumptions. I tend to agree with Albert Camus, who wrote in a footnote to *The Myth of Sisyphus* that "even the most rigorous epistemologies imply metaphysics. And to such a degree that the metaphysic of many contemporary thinkers [such as Rhees] consists in having nothing but an epistemology." Albert Camus, *The Myth of Sisyphus and Other Essays* (New York: Vintage International, 1991), 44.

54. Within Rhees' own framework, it is at least fair to say that Weil is asking the reader to consider that there is more than one way to think and talk about science and mathematics. When Rhees resists using spiritual and religious language to talk about science and mathematics, he is implying that we do not normally talk about science and mathematics in such a manner, hence to do so is incoherent. Weil is showing that it *is* possible to use such language with reference to science and mathematics. Doing so, as the Greeks did, opens up many fruitful possibilities for knowledge and understanding that are unavailable if we assume that such discourse about science and mathematics is inappropriate.

55. *NB*, 31.

56. Ibid., 546.

57. Ibid., 548.

58. Ibid., 440.

59. For an extended investigation of the importance of *metaxu* in Weil's thought, see Springsted's *Christus Mediator*, 197–219.

60. "The Romanesque Renaissance," *SE*, 46.

61. "God in Plato," *SN*, 90.

62. *SL*, 118.

63. "Philosophy," *FW*, 286.

64. *NB*, 453.

65. Ibid., 69.

66. Ibid., 370. It is also important to realize that the notion of bridges (*metaxu*) for Weil also includes participation in what the bridge is connecting. In other words, even though bridges must not be considered as ends in themselves but as leading to an end different from themselves, the bridges share in that end and, as such, cause us to share in it also.

67. "The Romanesque Renaissance," *SE*, 46.

68. Ibid., 47.

69. *NR*, 278. Weil's reference in the passage to mathematics as a "double language" is helpful in addressing the puzzle that arises from her claim that "eternal mathematics . . . is the stuff of which the order of the world is woven" (*NR*, 279; see note 14 above). Mathematics, as a double language, can be understood as both the thread of which the world is woven and as an integral part of the fabric thus woven.

70. *NB*, 514.

71. Dawkins, *Unweaving the Rainbow,* 63.

72. *NR*, 279.

73. "Classical Science and After," *SN,* 18.

74. *NB*, 514.

75. "Classical Science and After," *SN,* 21.

76. *NB*, 387.

77. Ibid., 511.

78. Ibid., 441.

79. Ibid.

80. "The Pythagorean Doctrine," *IC,* 189.

81. Rhees, 56.

82. *NB*, 193.

83. Ibid., 550.

84. "God in Plato," *SN,* 101.

85. "Reflections on the Right Use of School Studies with a View to the Love of God," *WG,* 106.

86. "Classical Science and After," *SN,* 41.

CHAPTER FOUR.   The Divine Poetry of Mathematics

1. "The Teaching of Mathematics," *SN,* 71–74.

2. Ibid., 71, 72.

3. Ibid., 72.

4. Ibid., 73.

5. "Classical Science and After," *SN,* 21.

6. Plato, *Gorgias* 507d, quoted in "The Pythagorean Doctrine," *IC,* 155.

7. *SL,* 118.

8. "Notes on Cleanthes, Pherecydes, Anaximander, and Philolaus," *SN,* 144.

9. *NR,* 233–34.

10. *FLN,* 85.

11. Ibid.

12. "The Pythagorean Doctrine," *IC,* 160.

13. *NB,* 512.

14. *FLN,* 80.

15. "The Pythagorean Doctrine," *IC,* 165.

16. *NB,* 161.

17. *SL,* 117.

18. "The Romanesque Renaissance," *SE,* 46.

19. *SL,* 125.

20. Ibid.

21. Ibid., 124.
22. Ibid., 125.
23. Ibid.
24. "Science and Perception in Descartes," *FW,* 33.
25. *NB,* 8.
26. "Science and Perception in Descartes," *FW,* 32.
27. "A Sketch of a History of Greek Science," *IC,* 202.
28. Ibid., 203.
29. *NB,* 162.
30. *SL,* 114.
31. "A Sketch of a History of Greek Science," *IC,* 203.
32. "Classical Science and After," *SN,* 41.
33. *NB,* 392.
34. "Classical Science and After," *SN,* 4.
35. *SL,* 114.
36. "The Pythagorean Doctrine," *IC,* 161.
37. *SL,* 114.
38. Ibid.
39. Ibid.
40. "A Sketch of a History of Greek Science," *IC,* 202–3.
41. Ibid., 203.
42. Ibid.
43. "The Pythagorean Doctrine," *IC,* 162.
44. *SL,* 121.
45. *NR,* 278.
46. Tradition is not consistent concerning which of Pythagoras' attributed discoveries prompted the sacrifice of a bull. Commentators usually say that it was the discovery that the circle is the locus of the apices of the right-angled triangles having the same hypotenuse (figure 4.7); Weil, however, occasionally indicates that it was the discovery that the circle is the locus of proportional means (see *NB* 505, *NB* 528, *SN* 21, for instance), a discovery illustrated by figure 4.8. Strictly speaking, as we shall see, the two discoveries amount to exactly the same thing, since they make the connection between circles and right-angled triangles which have already been shown to be the source of proportional means.
47. *NR,* 278.
48. Ibid.
49. "The Pythagorean Doctrine," *IC,* 191.
50. Ibid., 192.
51. *NB,* 512.
52. Ibid., 505.
53. "Divine Love in Creation," *IC,* 100.
54. "Classical Science and After," *SN,* 21.

55. "The Pythagorean Doctrine," *IC*, 159.

56. Ibid.

57. Ibid.

58. Ibid.

59. *NB*, 603.

60. "The Pythagorean Doctrine," *IC*, 162.

61. *SL*, 120.

62. "The Pythagorean Doctrine," *IC*, 161.

63. "Notes on Cleanthes, Pherecydes, Anaximander, and Philolaus," *SN*, 144.

64. *SL*, 121–22.

65. Ibid., 115.

66. Ibid.

67. "The Pythagorean Doctrine," *IC*, 162.

68. "The Teaching of Mathematics," *SN*, 73.

69. "The Pythagorean Doctrine," *IC*, 161.

70. *SL*, 119.

71. Ibid., 120–21.

72. Ibid., 120.

73. "The Pythagorean Doctrine," *IC*, 163.

74. Plato, *Epinomis* 990d, quoted in "The Pythagorean Doctrine," *IC*, 156.

75. *NB*, 392.

76. Ibid., 603.

77. *SL*, 116.

78. "A Sketch of a History of Greek Science," *IC*, 203.

79. Eric O. Springsted, "Contradiction, Mystery, and the Use of Words in Simone Weil," in *The Beauty That Saves,* ed. John M. Dunaway and Eric O. Springsted (Macon, Ga.: Mercer University Press, 1996), 16–17.

80. "Classical Science and After," *SN*, 20.

81. Ibid., 45.

82. "The Pythagorean Doctrine," *IC*, 164.

83. *NB*, 412.

84. Ibid., 162.

85. Ibid., 387.

86. Ibid., 514.

87. "The Pythagorean Doctrine," *IC*, 164.

88. *NB*, 412.

CHAPTER FIVE.   "All Geometry Proceeds from the Cross"

1. Czeslaw Milosz, *Emperor of the Earth: Modes of Eccentric Vision,* 89.

2. "Spiritual Autobiography," *WG*, 67.

3. Ibid., 68.

4. *WG*, 4–5. Fiedler's claim, although well taken, is undoubtedly an over-statement. The Apostle Paul, for instance, was actively hostile to Christianity, while Weil was at most reluctant. As the discussion below indicates, she believed after her mystical experiences that she had "always adopted to the Christian attitude as the only possible one" (*WG*, 62).

5. "Spiritual Autobiography," *WG*, 65.

6. Ibid., 68.

7. Ibid., 69.

8. "Variant: Meditation on Greek Science," *SN*, 47.

9. *NR*, 238.

10. "Spiritual Autobiography," *WG*, 69.

11. Springsted's *Christus Mediator* also treats this issue extensively, but does it in an entirely different manner than our present study. Springsted believes her understanding of Greek mathematics is subservient to her understanding of the cross. Hence, he establishes and discusses the Christian/theological nature of Weil's conception of mediation in the first half of his book, then turns to mediation in Platonic thought. Springsted makes it clear throughout his book that, in his estimation, although Weil's interest in Greek philosophy and mathematics precedes her marked movement toward Christianity, it is preferable to start with her understanding of Christ and the cross as mediator, then turn toward an investigation of the Greek/Platonic/Pythagorean elements.

I am inclined to believe that the procedure in our present study, to investigate Weil's conception of mathematics and science before considering how this conception illuminates her later thought after her mystical encounters, shines a different light on Weil's thought than is possible if one starts with the theological aspects of her thought. I am inclined to agree with what Springsted himself writes concerning this issue in *Christus Mediator*: "There is . . . a certain circularity in Weil's arguments concerning the mediation of the Cross and mediation in the Greeks. . . . But for all of this, the circularity is not vicious; rather, it serves to establish the consistency of the assumptions at various points" (139–40). It ultimately is fruitful to trace the relationship from the cross to mathematics (as Springsted does) *as well as* from mathematics to the cross (as I do), without necessarily committing oneself to which procedure has priority.

12. "The Pythagorean Doctrine," *IC*, 171.

13. *FLN*, 98.

14. *NB*, 441.

15. "Notes on Cleanthes, Pherecydes, Anaximander, and Philolaus," *SN*, 142.

16. Plutarch, *The Lives of the Noble Grecians and Romans*, trans. John Dryden, rev. Arthur Hugh Clough (New York: The Modern Library, 1970), 378.

17. Quoted by Weil in *FLN*, 25.

18. "A Sketch of a History of Greek Science," *IC*, 207.

19. Ibid.

20. Ibid.

21. "The Romanesque Renaissance," *SE,* 46.

22. *NB,* 454.

23. Ibid., 480.

24. *FLN,* 191, 194.

25. *NR,* 280–81.

26. *FLN,* 109.

27. Ibid.

28. Ibid.

29. "The Pythagorean Doctrine," *IC,* 170.

30. Ibid., 164.

31. *FLN,* 110.

32. "The Pythagorean Doctrine," *IC,* 197.

33. Ibid., 166.

34. Ibid., 196.

35. *FLN,* 80.

36. "Notes on Cleanthes, Pherecydes, Anaximander, and Philolaus," *SN,* 145.

37. Ibid., 144.

38. "The Love of God and Affliction," *SN,* 187.

39. Ibid., 176.

40. *NB,* 341.

41. Ibid.

42. Ibid.

43. "Notes on Cleanthes, Pherecydes, Anaximander, and Philolaus," *SN,* 141.

44. "The Love of God and Affliction," *SN,* 176–77.

45. "Divine Love in Creation," *IC,* 93.

46. *NR,* 278.

47. *NB,* 406.

48. "The Pythagorean Doctrine," *IC,* 167–68.

49. "The Love of God and Affliction," *SN,* 176.

50. "The Pythagorean Doctrine," *IC,* 168.

51. *FLN,* 120.

52. "The Love of God and Affliction," *SN,* 194.

53. Ibid., 176.

54. *NB,* 386.

55. "The Pythagorean Doctrine," *IC,* 154.

56. *NB,* 460.

57. Ibid., 587.

58. "Notes on Cleanthes, Pherecydes, Anaximander, and Philolaus," *SN,* 142–43.

59. Ibid., 143.

60. Allen and Springsted, *Spirit, Nature, and Community,* 44.
61. "The Pythagorean Doctrine," *IC,* 191–92.
62. Ibid., 168.
63. Ibid.
64. "Divine Love in Creation," *IC,* 95.
65. "The Pythagorean Doctrine, *IC,* 169.
66. Ibid.
67. Ibid.
68. Ibid.
69. *FLN,* 70.
70. Revelation 13:8, quoted in "Divine Love in Creation," *IC,* 93.
71. "The Love of God and Affliction," *SN,* 174–75.
72. Ibid., 176–77.
73. Ibid., 181.
74. "The 'Republic,'" *IC,* 142.
75. "The Love of God and Affliction," *SN,* 175.
76. *NB,* 388.
77. "The Pythagorean Doctrine," *IC,* 170.
78. Ibid., 172.
79. *NB,* 385.
80. *Epinomis* 990d, quoted in "The Pythagorean Doctrine," *IC,* 156.
81. *NB,* 385.
82. Ibid.
83. *NB,* 439.
84. "The Pythagorean Doctrine," *IC,* 161.
85. Ibid.
86. *NB,* 615.
87. "Notes on Cleanthes, Pherecydes, Anaximander, and Philolaus," *SN,* 144.
88. Ibid., 142.
89. "The Pythagorean Doctrine," *IC,* 161.
90. "The 'Republic,'" *IC,* 141.
91. *NB,* 385.
92. Weil sometimes suggests that there can be direct equality between humanity and God, as when in her London notebook she writes, as a gloss on Matthew 24:45–47, that "the reward is certainly a total identification with God" (*FLN,* 339). It is most consistent, however, with her discussions of Christ as mediator that whatever equality there can be between human and divine is only made possible via a mediator who is both human and divine (the Christ).
93. "The Pythagorean Doctrine," *IC,* 170.
94. Ibid., 171.
95. "The 'Republic,'" *IC,* 141.
96. Ibid., 138.

97. Ibid., 143.

98. *FLN,* 91.

99. "The 'Republic,'" *IC,* 142.

100. I Corinthians 15:17.

101. Ibid., 143.

102. "Reflections on the Right Use of School Studies with a View to the Love of God," WG, 112.

CHAPTER SIX.   "Our Country Is the Cross"

1. *NB,* 173.

2. "Forms of the Implicit Love of God," *WG,* 178.

3. "Classical Science and After," *SN,* 41.

4. "The Pythagorean Doctrine," *IC,* 196.

5. *GG,* 86.

6. *FLN,* 269–70.

7. "The Love of God and Affliction," *SN,* 187.

8. *FLN,* 43–44.

9. Ibid., 44.

10. "The Pythagorean Doctrine," *IC,* 172.

11. Ibid., 172.

12. Ibid., 173.

13. "Are We Struggling for Justice?", *SW,* 121.

14. "Forms of the Implicit Love of God," *WG,* 142.

15. "The Pythagorean Doctrine," *IC,* 173.

16. Ibid.

17. *FLN,* 87.

18. "Forms of the Implicit Love of God," *WG,* 142.

19. "Draft for a Statement of Human Obligations," *SE,* 220.

20. "The Pythagorean Doctrine," *IC,* 175.

21. *FLN,* 87.

22. Thucydides, *The Peloponnesian War,* trans. Rex Warner, introduction and notes by M. I. Finley (London: Penguin Books, 1974), 400–408.

23. "The Pythagorean Doctrine," *IC,* 174.

24. Friedrich Nietzsche, *The Will to Power,* trans. Walter Kaufmann (New York: Vintage Books, 1967), 234.

25. "Are We Struggling for Justice?", *SW,* 123.

26. Ibid.

27. Ibid.

28. "Forms of the Implicit Love of God," *WG,* 143.

29. "Are We Struggling for Justice?", *SW,* 122.

30. "The Pythagorean Doctrine," *IC*, 177.
31. "Are We Struggling for Justice?", *SW*, 122.
32. Ibid., 130.
33. Ibid., 123.
34. Ibid., 128.
35. "Draft for a Statement of Human Obligations," *SE*, 219. Although Weil affirms that human beings have this power to turn toward an absolute good in a number of texts, she also says in "Forms of the Implicit Love of God" (*WG*, 211) that "They do not turn toward God. How could they when they are in total darkness?", and in other texts she frequently stresses that we cannot go out to find God, but must wait for God to come to us. This apparent contradiction can be resolved by realizing that we do not save ourselves, but can only consent to God's action. The human being does have the power to consent or to not consent. This power to consent is precisely that "power of turning" that Weil is referring to.
36. Ibid., 220.
37. Ibid.
38. Ibid.
39. *NB*, 274.
40. "Draft for a Statement of Human Obligations," *SE*, 221–22.
41. "Forms of the Implicit Love of God," *WG*, 146.
42. Ibid., 149.
43. In other texts, (*WG*, 98 and 151, for instance), Weil equally stresses the impersonal and universal aspects of true friendship.
44. Ibid., 148.
45. Ibid., 144.
46. Ibid.
47. Ibid., 208.
48. "The Pythagorean Doctrine," *IC*, 177.
49. Ibid., 191.
50. Ibid., 185.
51. "Forms of the Implicit Love of God," *WG*, 159–60.
52. There are a number of fine secondary studies of necessity in Weil. See, for instance, Allen and Springsted, "Divine Necessity: Weilian and Platonic Conceptions," in *Spirit, Nature, and Community* (33–52), as well as Rhees' discussion in *Discussions of Simone Weil*, 54–64.
53. "The Love of God and Affliction," *SN*, 195.
54. "The Pythagorean Doctrine," *IC*, 184.
55. Ibid., 177.
56. Ibid., 178.
57. Ibid., 185.
58. *NB*, 274–75.
59. "The Pythagorean Doctrine," *IC*, 186.

60. Ibid., 179.

61. Ibid., 178.

62. Ibid., 182.

63. *NR*, 277.

64. "The Pythagorean Doctrine," *IC*, 181.

65. Ibid., 187–88. Weil may be overstating her case here when she says that the "necessary connections" only have reality as an object of intellectual attention. We have seen frequently that she believes, in a Kantian fashion, that the human being is in a real sense a "co-creator" of the world. The reality of the world, however, does not depend on us. When we co-create the world in love, it is to make it fully real to us; but it is already real.

66. Ibid., 182.

67. Ibid., 188.

68. *NR*, 279.

69. "The Pythagorean Doctrine," *IC*, 188.

70. Ibid., 180.

71. Ibid., 184.

72. *FLN*, 80.

73. "The Love of God and Affliction," *SN*, 186.

74. Ibid., 175.

75. "Forms of the Implicit Love of God," *WG*, 177. There is, at least, no overall natural worldly purpose to these laws and relationships. As we shall see, however, they do have a supernatural purpose.

76. Ibid.

77. "The Love of God and Affliction," *SN*, 193.

78. "The Pythagorean Doctrine," *IC*, 198. This passage identifies at least part of the supernatural purpose referred to in note 75 above.

79. *NR*, 251.

80. "The Pythagorean Doctrine," *IC*, 184.

81. *NR*, 269.

82. Ibid., 259.

83. Ibid., 269.

84. Ibid., 272.

85. Ibid., 270.

86. "The Love of God and Affliction," *SN*, 187.

87. "Forms of the Implicit Love of God," *WG*, 159.

88. *NR*, 272.

89. Ibid., 275, 277.

90. "The Love of God and Affliction," *SN*, 193.

91. *NR*, 275.

92. "The Love of God and Affliction," *SN*, 179.

93. Ibid., 186.

94. Ibid., 178.
95. "The Pythagorean Doctrine," *IC*, 187.
96. Ibid., 182–83.
97. Job 13:15.
98. "The Pythagorean Doctrine," *IC*, 187.
99. Ibid., 193, 195.
100. "The Love of God and Affliction," *SN*, 186.
101. Dawkins, *Unweaving the Rainbow*, 63.
102. "Forms of the Implicit Love of God," *WG*, 162–63.
103. "The Pythagorean Doctrine," *IC*, 196.
104. "Forms of the Implicit Love of God," *WG*, 164–65.
105. *FLN*, 329.
106. "Forms of the Implicit Love of God," *WG*, 164.
107. "The Love of God and Affliction," *SN*, 180.
108. "The Pythagorean Doctrine," *IC*, 189.
109. *NB*, 241.
110. "Forms of the Implicit Love of God," *WG*, 163, 171.
111. Ibid., 164.
112. *FLN*, 98.
113. "The Pythagorean Doctrine," *IC*, 191.
114. "The Love of God and Affliction," *SN*, 198.
115. "Forms of the Implicit Love of God," *WG*, 175.

CONCLUSION.   "A Preparation for Deliverance"

1. *NB*, 441.
2. *NR*, 281.
3. "The Pythagorean Doctrine," *IC*, 191.
4. *NR*, 227.
5. *SL*, 91.
6. "Reflections on Quantum Theory," *SN*, 64.
7. "Fragment: Foundation of a New Science," *SN*, 79.
8. Ibid., 80.
9. Ibid., 79.
10. Ibid., 80.
11. Ibid., 79.
12. Ibid., 80.
13. Ibid., 81.
14. "Forms of the Implicit Love of God," *WG*, 164.
15. *NR*, 250.
16. Ibid., 235.

17. Ibid., 269–70.
18. Ibid., 235.
19. Ibid., 249.
20. "Reflections on Quantum Theory," *SN*, 62.
21. *NR*, 243.
22. Ibid., 242.
23. Ibid., 281.
24. Ibid., 242.
25. "The Love of God and Affliction," *SN*, 194.

Bibliography

WORKS BY SIMONE WEIL

First and Last Notebooks. Translated by Richard Rees. London: Oxford University Press, 1970.

Formative Writings, 1929–1941. Edited and translated by Dorothy Tuck McFarland and Wilhelmina Van Ness. Amherst: University of Massachusetts Press, 1988.

Gravity and Grace. Translated by Emma Craufurd. London: Routledge & Kegan Paul, 1987.

Intimations of Christianity Among the Ancient Greeks. Translated by E. C. Geissbuhler. London: Routledge & Kegan Paul, 1957. All references include reference to a specific essay.

Lectures on Philosophy. Translated by Hugh Price, introduction by Peter Winch. Cambridge: Cambridge University Press, 1978.

The Need for Roots. Translated by Arthur Wills. London: Routledge & Kegan Paul, 1952.

The Notebooks of Simone Weil, 2 volumes. Translated by Arthur Wills. London: Routledge & Kegan Paul, 1956.

On Science, Necessity, and the Love of God. Translated by Richard Rees. London: Oxford University Press, 1968. All references include reference to a specific essay.

Selected Essays, 1934–43. Chosen and translated by Richard Rees. London: Oxford University Press, 1962. All references include reference to a specific essay.

Seventy Letters. Translated by Richard Rees. London: Oxford University Press, 1965.

Simone Weil. Edited by Eric O. Springsted. Maryknoll, N.Y.: Orbis Books, 1998.

Waiting for God. Translated by Emma Craufurd. New York: Harper and Row, 1973. All references include reference to a specific letter or essay.

SECONDARY WORKS CITED

Allen, Diogenes, and Eric O. Springsted. *Spirit, Nature, and Community*. Albany: State University of New York Press, 1994.

Aristotle. *The Basic Works of Aristotle*. Edited and translated by Richard McKeon. New York: Random House, 1941.

Camus, Albert. *The Myth of Sisyphus and Other Essays*. Translated by Justin O'Brien. New York: Vintage International, 1991.

Dawkins, Richard. *Unweaving the Rainbow*. Boston: Mariner Books, 1998.

Descartes, René. *Oeuvres de Descartes*. Edited by Charles Adam and Paul Tannery. 12 volumes. Paris: Vrin, 1897–1910.

Guthrie, Kenneth Sylvan. *The Pythagorean Sourcebook and Library*. Introduced and edited by David R. Fideler. Grand Rapids: Phanes Press, 1987.

Lassere, François. *The Birth of Mathematics in the Age of Plato*. Cleveland: Meridian Books, 1966.

McKirahan, Jr., Richard D. *Philosophy Before Socrates*. Indianapolis: Hackett Publishing Company, 1994.

Milosz, Czeslaw. *Emperor of the Earth: Modes of Eccentric Vision*. Berkeley: University of California Press, 1977.

Murdoch, Iris. *The Sovereignty of Good*. London: Routledge, 1970.

Nietzsche, Friedrich. *The Will to Power*. Translated by Walter Kaufmann. New York: Vintage Books, 1967.

Planck, Max. *Initiations à la physique*. Paris: Flammarion, 1941.

Plutarch. *The Lives of the Noble Grecians and Romans*. Translated by John Dryden, translation revised by Arthur Hugh Clough. New York: The Modern Library, 1970.

Rhees, Rush. *Discussions of Simone Weil*. Edited by D. Z. Phillips. Albany: SUNY Press, 2000.

Springsted, Eric O. *Christus Mediator: Platonic Mediation in the Thought of Simone Weil*. Chico, Calif.: Scholars Press, 1983.

———. "Contradiction, Mystery, and the Use of Words in Simone Weil." In *The Beauty That Saves*, edited by John M. Dunaway and Eric O. Springsted. Macon, Ga.: Mercer University Press, 1996.

———. "Spiritual Apprenticeship." In *Cahiers Simone Weil* 25 (2002).

Thucydides. *The Peloponnesian War*. Translated by Rex Warner, introduction and notes by M. I. Finley. London: Penguin Books, 1974.

Winch, Peter. *Simone Weil: "The Just Balance."* Cambridge: Cambridge University Press, 1989.

Zukav, Gary. *The Dancing Wu Li Masters*. New York: Bantam Books, 1980.

Index

Pythagoreanism (*cont.*)
and limit, 73–74
and mathematics, 71–72, 83, 95, 98,
110, 113–144, 120–21, 125, 135, 149
and metaphysics, 68–69, 72–73,
79–80, 86, 88, 98, 100, 110, 114–15,
119–20
and music, 74–79
and number, 69, 71–75, 114–15, 118,
120, 134, 150, 162
and Plato, 69, 86, 153–54
and prophecy, 122, 129–30
and proportion, 71, 79, 83, 98, 107, 110,
114–15, 118, 124, 143, 150, 154
and religion, 68–69, 83, 98, 100, 110
and science, 69, 83, 100
and the Trinity, 138–39

quanta, 56
quantum theory, 7, 11, 47–58. *See also*
science, contemporary
and contradiction, 54–55, 62, 125
counterintuitive aspects of, 48–51,
61–62
and discontinuity, 53–56
and energy, 49–51, 56–57, 59, 136
success of, 47, 52, 64

ratio (*logos*), 79, 102, 104–5, 114, 125
irrational (*logoi alogoi*), 120, 123, 125
reason, 18–20, 30, 61–62, 126, 133–34, 140
relativity, theories of, 7, 11, 47–48, 53
counterintuitive aspects of, 48, 62
religion
and art, 5, 10, 33, 69, 80, 85, 88
and mathematics, 68–69, 83, 98, 101,
110, 147
and philosophy, 84, 86, 88
and Pythagoreanism, 68–69, 83, 98,
100, 110
and science, 3, 7, 10, 64, 69, 80, 85,
88, 100, 130–31, 178, 188, 191–94
and truth, 86, 193
and Weil, 24, 127–29, 192–94
Renaissance, 87, 98
Republic (Plato), 154

respect, 166–68
Resurrection, 154–55
Reynaud-Guérithault, Anne, 12
Rhees, Rush, 81–89
admiration for Weil, 81–84, 85
critique of Weil, 81–84, 87, 94
and Pascal, 86–87
and Wittgenstein, 81, 84
Rules for the Direction of the Mind
(Descartes), 20

sacrament, 184
mathematics and science as, 2, 4–6,
130–31
sacrifice, 159, 180
of Christ, 154–55
and creation, 147, 182
and Pythagoras, 110–11, 114
of the Son, 145–48
science, 27, 106. *See also* science,
classical/modern; science,
contemporary; science, Greek
and art, 3, 5, 10, 68–69, 80, 85, 89
authority of, 14, 22, 37–38, 63–64
and beauty, 3, 8–9, 46, 82–83, 88, 91,
93, 158, 190–91, 194
as bridge (metaxu), 90–91, 187
and common sense, 14–17, 48, 59, 61
and contradiction, 45–47, 53–54
definitions of, 3–6, 33, 187–89
and God, 4, 6, 68, 129, 132, 189–93
limits of, 4, 64–65, 189–90
and love, 147, 175, 186, 187–91,
193–94
and metaphysics, 19, 21, 48, 86–90,
100, 190–91
and morality, 34, 42, 82, 158
and necessity, 171, 173–74, 176
as obedience, 5, 180
object of, 3–5, 18, 187–88
and order, 158, 194
origins of, 13–14
and philosophy, 2, 60, 62, 80–82,
86, 88
as power, 4, 7–10, 33–35, 38, 63, 65,
68, 88, 132, 170, 176, 188–91, 194

VANCE G. MORGAN

is professor of philosophy at Providence College.